RISINGTIDEFALLINGSTAR

RISINGTIDEFALLINGSTAR

Philip Hoare

FOURTH ESTATE · London

For Pat

Chapter initial illustrations by Joe Lyward

'The sea, everywhere the sea, and no one looking at it'
DANY LAFERRIÈRE

THERISINGSEA

Not long ago but long enough, I looked into the old cupboard in my bedroom and at the back, among the piles of floppy discs and peeling spines of my children's encyclopaedias, I found a notebook. It was in an old-fashioned imperial format, half-bound with blue cloth and shiny paper, its fore-edge delicately spattered like a blackbird's egg. It came from the cable factory where my father had worked all his life. Inside, on feint-lined pages intended for notes on amps and electrical resistance, were writings and drawings I'd done when I was about fifteen years old.

On each left-hand page was a picture, in bright poster paint: a futuristic city, art deco designs, lithe figures out of some space opera or Russian ballet; fantastical images I'd collected in my teenage head. Halfway through the book I'd painted something I'd really seen: a leaping killer whale, slick with clear nail varnish to mimic its black-and-white skin, as if it had jumped out of the sea, rather than a concrete pool in a suburban safari park.

On the right-hand pages I'd composed lyrics and prose, the things I couldn't say out loud. Looking at this parade of longings forty years later, I realised that the fifteen-year-old me had mapped out his life along those pale blue lines. As if I'd already lived in reverse. Everything that came after had been entered in that blue notebook, balanced on my knees while I watched television in our front room, waiting for whatever might come next.

The wind howled at my window like a wild animal, a snarling beast demanding to be fed. The house held fast against horizontal rain that threatened to find every crack in the walls. The air was full of water, driven directly from the shore. Between the falling trees and the pounding waves, it seemed that the sea – for all that it was a mile away or more – was reaching out for me in the darkness. The newspapers and the television and the websites warned us not to walk near it, as if our mere approach might be dangerous, as if its tentacles might reach out and drag us in.

Growling and yowling, ranting and rocking, falling back to catch their breath before their next assault, the storms kept on coming, and there was nothing we could do. The world had become turbulent with its own temper, its air sweeping over oceans in a tropical fury. If we ever felt guilty, we felt it now.

At least the sea is visible in its rage; the wind is an unseen monster. You don't hear the wind; you hear what it leaves behind. It is defined by what gets in its way – trees, buildings, waves. Perhaps that's why it preys on our imagination so disturbingly. The spinning of the globe seemed to have become audible – the sound of a world out of kilter. For what sins were we being punished? What had we done wrong? In Caribbean hurricanes during the seventeenth century, Spanish priests would toss crucifixes into the waves or hold the Host up into the wind, for fear that their sinful flocks were responsible for God's displeasure.

That winter, storm after storm raked southern England. Tearing and snapping, the wind never seemed to stop. As I lay in bed, I could feel its volume whipping and squalling around me, changing direction wilfully, a mad car out of control.

Then, just when it seemed it could not get any worse, a mighty gale, as near as we might get to a hurricane, ripped out of the cover of night and into the naked day. Unable to sleep, disturbed by the charged air, as if its ions were crackling in my brain, I cycled down to the shore and took shelter under the eaves of the yacht club, a wooden building which seemed about to whirl off into the wind. Behind me stood a medieval abbey, and a fort once visited by the Virgin Queen to survey her maritime kingdom, its Tudor ramparts now protected from the waves by a long sea wall.

I've known this shore all my life: from its ancient Seaweed Hut – a weird structure which might as well have been put up by Iron Age inhabitants – to the brutal towers of its nineteen-sixties housing estate. It is as familiar to me as it is to the birds that scrabble for their livelihoods in its shingle and mud. I'd taken it for granted, that it would always be there.

I couldn't believe what I was seeing. The beach was being torn apart before my eyes. The wall, usually only lapped even at the highest spring tide, was entirely overwhelmed. Waves – to call them waves seems pathetically inadequate – had lost their laterality and gone vertical, rising higher than a house.

My world had lost its moorings. This was not some rocky Cornish or Scottish coast, buttressed against such a battering; this was a sedate, suburban shore, complacent and unprepared; a soft place on the southern edge of England, open to the rest of the world, successively invaded and settled for millennia. This estuary even had its own Roman deity, Ancasta. Clearly, she had been offended.

It was as though someone had computer-generated the weather and ramped it up to a ridiculous degree. An invisible alien, formed of roaring air and raging water, had been unleashed. The sea spray reached the tops of the trees on the shore. It was terrifying, and exhilarating. My heart raced to keep up with every rattling rolling rumble; a cacophony created by raked-up shingle and creaking trees, the Foley effects of enraged gods flinging nature around.

I watched it like some viral video; not rerun, but in real time. Behind this frontline, people were driving cars, taking buses, going to work, school, shops, locked in their own personal climate. We shared the same city; but they felt safe, seeing the storm through their screens. I was on the edge of it, physically confronted by the violence, as shocking as if I'd come across a fist fight on the street.

The sea wall had been replaced by a wall of sea. The placid site where I propped up my bike every morning, where I'd leave my clothes and slip into the water, joining rather than entering it, had become a deadly, repulsive place.

It was the only day during those storms that I did not, could not swim; perhaps the only day that year. Even at the height of the past days' disruption I'd launched myself into the madness, defying the warnings. So what if anything went wrong? I didn't take my mobile phone in case of emergency because I don't have one. People say I should be careful; but why be careful, when we are so full of cares? This was the opposite of that. I glorified in my stupidity. Foolhardy, a hardy fool. I had rocked with the waves, holding my head above water like a shipwrecked dog, dodging planks and plastic buckets. A single trainer had floated past, then a motorbike helmet; I wondered if the head might still be in it. I was borne up by the rollercoaster ride, exultant and excited, although I had soon found myself spat back onshore.

Not that day. That day I had to admit defeat, and defer to a greater power.

During the night the wind woke me again, prowling around the house like a midnight demon, ready to suck me out of the window. The sound was beyond sound: one white noise comprised of many others, fit to eviscerate my dreams.

In the morning, not quite believing what had happened during the darkness – was that last night, or the night before; did I even imagine it? – I ventured out on the third day of the storm, expecting to see a newly devastated world.

But the streets looked the same, just as they do when you come back from holiday. Only a few fallen branches from the trees hinted at the mayhem of the small hours. I rode on down to the beach, not knowing what to expect, but expecting it anyway.

There I realised that the storm had taken its final revenge. Defeated by what it could do inland, it had reshaped the coast itself.

The beach had been lifted up and thrown back, creating a shingle tsunami. The path had been replaced by a tangle of branches and rope, a twisted mass of line and grass torn from some other shore, in the way drowned men's pockets are turned inside out. The flotsam lay still, but contorted with the torque and tension of the wind and water. Tiny balls of coloured plastic, like the roe of some new petrochemical sea creature, were scattered through the wrackline. The calm itself was violent.

Then I saw the sea wall. The waves had fallen back to reveal their guilty secret. The long straight stalwart of my pre-dawn swims, my chilly changing place, my launching point from the land, had been smashed to pieces, kicked over by a petulant child. The wall had stood for seventy years or more, made of the same stone that had built the abbey and fort behind it. Now, like the abbey, it lay in ruins.

I took it personally. A structure I knew as well as my own body had been reduced to rubble. And I knew I was responsible. I had allowed this to happen.

No one would ever rebuild this place, this insignificant corner, bypassed by ships and cars. The aftermath of this assault was the reality of 'managed retreat', in the bureaucrats' parlance; the desertion of an already forgotten site. We had abandoned beauty, abandoned nature. This was the future: the rising sea on a suburban shore. I wanted to cry, but the hardness of the stone stopped me. So I picked my way over the remains, pulled off my clothes, and got in.

The sea was still filled with debris; household doors and tree trunks floated like lumber thrown from a giant's outhouse. And even as I swam, the waves began to rise again, responding to an invisible moon. Giving up the struggle, I climbed out, fighting to get dressed as the wind whipped my clothes into air-filled versions of me.

I rode off into a new landscape. Time had sped up; geological change happened overnight. New streams had been formed and new islands created in the flood. Reluctant to let me go, the waves lashed me as I passed.

Frail as I am as a human, I had the ability to withstand the storm. But along this coast, thousands of seabirds died in those few days, ten thousand guillemots alone. Birds that mate for life waited for partners who would never return.

That afternoon I found a dead guillemot on the beach. Its slender-sharp bill and black-brown body lay slumped on the shingle like a soft toy lost over the side of a passing ferry. It was so perfect – and so far from the rocky ledges where it breeds so close to its fellows that they preen one another, even though they are not partners – that I wondered if I should take it home with me. I talked to it, commiserating with its fate. The next tide claimed it – only to bring another victim, rolled about in nylon line.

I pulled it out. It was an avocet. A delicate, emblematic bird I'd only ever seen from afar, suddenly brought into near focus, graphically black and white, a piece of netsuke. I don't think I've ever seen anything so exquisite, there in my hands. I turned it over in my fingers, feeling its long, scaly legs – relics of its reptilian past – and their knobbly joints like threaded veins, or worms that had swallowed soil. It was their colour that amazed me: an indefinable, pearly blue-mauve withheld under a soft misty bloom; electric, verging on iridescent. I could only compare it to the oceanic blue-grey of a gannet's bill, as if this were a marine colour reserved for a seabird's exclusive use. The legs, which might have been made out of some alien metal or art nouveau glass, culminated in tiny black claws, embedded like chips of polished jet.

It was an enamelled animal. Vitrified. As much jewellery as a live, or recently dead, thing. It was hard to believe it had stood on such fragile stilts, let alone stalked its prey from them. Then I remembered the living avocets I'd seen, moving with eighteenth-century elegance, as if they might dance a gavotte. Avocets enact their own rituals, gathering in a circle and bowing to one another like dandies.

I opened the bird's wings, a pair of fans fluttering across a ballroom. They felt taut with what they were expected to do; not yet quite useless, yet no use now, either. They articulated lightness and lift. The black-capped head, which once bobbed in the shallows, sweeping through the water with its upturned bill, ended in the final, defining, typographical tick that is the avocet's glory. It was almost too beautiful to touch, but I prised open the ebony splint, like the split reed of a musical instrument. Its lower half was precisely ridged to speed its ploughing; a keratin tool engineered to micron perfection, tapering to a paper-thin tip, as sharp as a squid's beak. I remembered the sound that played through it, a shapely insistent peep, accompanying the nervously graceful movement as the creature swung its bill from side to side in search of invertebrates. Even the bird's binomial expressed its exotic allure – *Recurvirostra avosetta*, as if it were a minor Egyptian god.

With a wrench and a twist, I pulled off the head. The muscles and oesophagus came away, dangling raw and red. Then I spread the body on a long piece of flotsam, laying out the bird on the knotted wood, under the grey sky.

Beautiful, but broken.

———

As the year slows to its midnight, the solstice blows in fierce and wild, the last of December putting up a fight. Day and night blur; it's difficult to say when one becomes the other. In the glittering darkness long before dawn, a silver ring of ice is slung around the moon, catching stars and planets in its circle. Their heavenly bodies hang in the O: orbits within orbits, eyes within eyes. I swim through the inky sea, my white body breaking the black surface, moving through the moon.

The tide is high again. It often is here, more than most places, since Southampton Water experiences an unusual double tide, standing twice as high and twice as low every day, swelled and drained by the Atlantic Pulse that drives up and down the Channel. In *David Copperfield*, Dickens's watery, autobiographical book, Mr Peggotty, mindful of his 'drowndead' relations, says of Mr Barkis, 'People can't die, along the coast . . . except when the tide's pretty nigh out. They can't be born, unless it's pretty nigh in – not properly born, till flood. He's a going out with the tide.' The rise and fall brings life and death beyond our control. When the moon is full, the tide is high at noon and midnight, like a clock. The tide is time; the two words share the same root, as does tidy. We are all tidied up by time.

I stand on what is left of the sea wall in the moonlight, charged by its brightness. Apparently Siberian shamans would strip naked during the full moon to absorb its energy; maybe it'll warm me with its secondhand daylight. The satellite silences our world; it has mysterious powers, as Bernd Brunner notes, still not quite explained, like black holes or gravity itself: some scientists believe the lunar effect extends to the land too, triggering earthquakes as though the planet's tectonic slides were tides of their own.

And if our home is a living thing, then the sea is its pumping heart, swelling as the moon swings around the earth, tugging at our blood, at the tide inside of me. After all, the entire planet consists mostly of water, like us, and we are governed by its cycles more powerfully than by any elected body. Its tides are our future.

They are always racing ahead, every day an hour further on, a re-minder that we will never catch up with ourselves, no matter how fast we may swim.

But then, for me every day is an anxiety in my ways of getting to the water. I worry that something will stop me from reaching it, or that one day it won't be there – as it is, and it isn't, twice a day. I've become so attuned to it, so scared of it, so in love with it that sometimes I think I can only think by the sea. It is the only place I feel at home, because it is so far away from home. It is the only place where I feel free and alive, yet I am shackled to it and it could easily take my life one day, should it choose to do so. It is liberating and transforming, physical and metaphysical. Without its energy, we would not exist. There is nothing so vast in our lives, so beyond our temporal power. If there were no oceans, would we have our souls? 'The sea has many voices | Many gods and many voices,' T. S. Eliot wrote. 'We cannot think of a time that is ocean-less.' 'In civilisations without boats,' said Michel Foucault, 'dreams dry up.' Even if we could live without the oceans, a world of arid plains and dry valleys would lack mystery; everything would seem knowable, exposed.

In the womb we swim in salty water, sprouting residual fins and tails and rudimentary gills as we twist and turn in our little oceans. It was a tradition in maritime communities that if a child was born with the amniotic sac, the caul, over its head, she or he would never drown, having survived this near-suffocation. To be born thus was to be 'born behind the veil', and a preserved caul – itself a veil between life and death – would extend protection to anyone who carried it: David Copperfield is born with a caul which is auctioned when he is ten years old, leaving him uncomfortable and confused at having part of himself sold off. We first sense the world through that fluid filling our mother's belly; we hear through the sea inside her. The sea is an extension of ourselves. We speak of bodies of water, and Herman Melville wrote of 'the times of dreamy quietude, when beholding the tranquil beauty and brilliancy of the ocean's skin'. Compared to the thin epidermis of land we occupy, the great volume of the sea exceeds our sway; it lends our planet its depth, and ourselves a sense of depth.

And if we are mostly water, hardly here at all, then other celestial bodies might be entirely aquatic. An astrophysicist once told me about newly discovered exoplanets that may be composed of water hundreds of kilometres deep, with only a few rocks at their hard core. Disdaining our need for land, these globular oceans, spinning translucently in some distant galaxy, may be inhabited, as astrobiologists hypothesise – it being their business to study that which may or may not exist – by giant whale-like creatures, half-swimming, half-flying through their atmospheres.

The ubiquity of the sea – from this grey estuary in which I swim, to the great open oceans – is itself interplanetary, connecting us to the stars, not really part of our world at all. It doesn't begin until it begins, and then it never seems to end. It writes itself in the clouds and the currents, a permanently changing script, inscribing and erasing its own history, held down by air and gravity in a tacit agreement between land and sky, filling the space in between. It's a nothingness full of life, home to ninety per cent of the earth's biomass, providing sixty per cent of the oxygen we breathe. It is our life-support system, our greater womb. It is forever breaking its own boundaries, always giving and always taking. It is the embodiment of all our paradoxes. Without it we couldn't live, within it we would die. The sea doesn't care.

Down there lies another history, the unseen record of what is going on up above. Preserved in the freezing vaults of the National Oceanography Centre in Southampton are sample cores from the sea bed, long columns of mud and sediment whose layers tell out deep time like the rings in a tree or the waxy plugs in a whale's ear. Composed of falls of marine snow – minute animals and plants and minerals, the makings of limestone- and chalk-to-be – along with dark strata deposited by ancient tsunamis, their past is our future foretold. The water itself has an age, up to four thousand years old, a story of its own. And even if the sea has become a carbon sink, absorbing the energy we have released from the sun, this cistern of our sins is still the repository of our dreams.

But as I just told you, the sea doesn't care. It deals life and death for innocent and guilty alike.

The Tempest, Shakespeare's last and most watery play, was first performed at court for James I on All Saints' Day, 1611. It opens uproariously, slapping the audience in the face with a life-threatening storm and 'fraughting souls' on a ship about to split. In the dramatic tumult, panic spreads blame. Antonio, the usurping Duke of Milan, curses the boatswain – who is trying to save the ship – as a 'wide-chopped rascal – would thou mightst lie drowning | The washing of ten tides!' He is arrogantly invoking the practice of hanging pirates on the shore, leaving their corpses to swing in successive tides: 'He that's born to be hanged need fear no drowning.'

Yet, as the audience slowly becomes aware, these scenes of rip, wreck and panic – overturning all order as the crew fight for their lives and the aristocrats' status counts for nothing in the face of the waves: 'What cares these roarers for the name of king?' – turn out to be nothing more than a magic trick, a theatrical effect within a theatrical effect, a storm raised by a sorcerer's art and his impish familiar. As Ferdinand, the king's son, his hair up-staring, leaps from the sinking vessel set aflame by Ariel's divided fire, he cries, 'Hell is empty | And all the devils are here.' (It is an image which may have been inspired by James I himself, author of Demonologie and personal supervisor of the torture of witches, who believed that during a voyage back from Oslo in 1590 his ship had been beset by storms summoned by witchcraft, and demons had been sent to climb its keel.)

Suddenly, and as if in a dream, the castaways find themselves in an eerie calm, on an island full of strange noises, peopled by beings they cannot quite discern; an alien place, although the survivors themselves are aliens too. Some of its spirits are only rumoured, like Sycorax the witch, named after sys for sow, and korax for raven, a fated bird from an 'unwholesome fen'. Others are all too present, like her son Caliban, a bastard creation, 'a savage and deformed slave', amphibious, half-man, half-fish, 'Legged like a man! And his fins like arms!' He is a chimeric creature, as if slithering out of an evolutionary sea; his counterpart is Ariel, an ambivalent, fluid spirit of the air who eludes definition and can be anywhere in an instant. Both are ruled over by the all-powerful magician Prospero in his water-bound exile.

Recently, on a shelf of stranded books being sold to benefit a bird sanctuary overlooking the Solent, I discovered a 1968 Penguin edition of the play. It was an oddly apt place to find it: this silted-up seventeenth-century harbour, overflown by marsh harriers and stalked by godwits and avocets, was the domain of the Earl of Southampton – Harry Southampton, Shakespeare's Fair Youth and possibly his lover, who lived at nearby Titchfield Abbey, where the playwright's works were performed.

I paid fifty pence for the book, attracted by its cover, designed by David Gentleman. Splashed with broad swathes of solid colour, the wood engraving, inspired by the work of Thomas Bewick, seemed to span the turbulent year of its nineteen-sixties publication – when protestors lifted up pavement stones, to find the beach below – and the uncertainties of its seventeenth-century contents.

A three-masted ship tilts in a stylised sea, rolling on waves below stormy clouds towards a tree-blown island and a rocky cave, all rendered in sludgy, overlapping shades of subdued blue and green, grey and teal, like the birds and land and sea around the building where I'd bought the book. The design was almost cartoon-like, folkloric, and layered. It caught the dark mystery and music of the words within.

The Tempest is a ceremony, a ritual in itself, publicly performed in a sky-open theatre on the site of the Blackfriars monastery on the Thames, a river into which sacrifices were once thrown to propitiate the gods. It is a pared-back, mysterious work, 'deliberately enigmatic', as Anne Righter says in her introduction to the Penguin edition, 'an extraordinarily secretive work of art', so emblematic that it might be acted out in mime, without any words at all.

Its origins lay in the fate of *Sea Venture*, which sank off Bermuda in 1609 while carrying colonists from Plymouth to Jamestown; Harry Southampton himself was an investor in the Virginian settlement. Shakespeare drew on William Strachey's account of the wreck, a natural history of disaster, with its tales of St Elmo's fire at the height of the storm – 'an apparition of a little round light, like a faint Starre, trembling, and streaming along with a sparkling blaze' – and the eerie calls of petrels coming in to roost, 'a strange hollow and harsh howling'. Their cries earned Bermuda its reputation as an island of devils, one which Strachey rationally dismissed, although he did acknowledge the presence of other monsters: 'I forbear to speak what a sort of whales we have seen hard aboard the shore.'

The Tempest is the closest Shakespeare comes to the New World. It is almost an American play, although two centuries later its castaways might have been washed up on another colony: Van Diemen's Land, on whose remote south-western shores one can still imagine a seventeenth-century shipwreck and its stranded sailors stumbling about on the alien sand. Some saw Caliban and Ariel as symbolic representations of newly-discovered native peoples, whose countries were already being plundered by the West; others have seen a reflection of an island nearer to home: Ireland, a troublesome place filled with its own wild people, and regarded as a plantation to be conquered. But equally, Prospero's isle might be utopia, a nowhere place over which his magic rises as a mist veiling time and space – just as a century earlier, Columbus, researching his expedition, had written notes and marginalia about strange people cast up on the shores of the Azores and the west of Ireland: 'We have seen many notable things and especially in Galway, in Ireland, a man and a woman with miraculous form, pushed along by the storm on two logs.'

Shakespeare, nearing the end of his life, appeared to have recreated himself in the omniscient magician; others have seen Prospero as a reflection of Elizabeth I's astrologer, John Dee, who communed with angels using a golden disc, and peered into his black obsidian mirror – stolen from the New World – in order to see the future and the past. The whole play seems to be happening before it was written. It is fraught in the original sense of the word, as a ship filled with freight, as well as with meaning. Shakespeare was familiar with the ocean: he refers to it more than two hundred times in his works, and some critics believe that he was once a sailor. Certainly he knew its meaning, and set The Tempest on a 'never-surfeited sea', a transformative place. After the storm, Ariel tells Ferdinand that his father, the king, lies 'full fathom five'; he has been made immortal by the water, becoming a baroque jewel in the process:

Of his bones are coral made;
Those are pearls that were his eyes;
Nothing of him that doth fade,

But doth suffer a sea-change
Into something rich and strange.

For the artists and poets who came after, The Tempest lingered in its magical power and deceptive simplicity. Samuel Taylor Coleridge thought that Prospero's art 'could not only call up the spirits of the deep, but the characters as they were and are and will be', and that Ariel was 'neither born of heaven, nor of earth; but, as it were, between both'. For Percy Shelley, who would be nicknamed Ariel, the play evoked 'The murmuring of summer seas' and his own in-between state. And for John Keats, in whose volumes of Shakespeare's works the play was the most heavily scored, it became a pattern for his imaginative life, like a map to be followed. Indeed, he sailed down Southampton Water with The Tempest in his pocket.

In April 1817, Keats, then a medical student in London, took the coach to Southampton in search of distraction. He had loved the sea ever since he'd read of 'sea-shouldering whales' in Spenser's The Faerie Queene – 'What an image that is!'; his poetry habit had made him 'a Leviathan...all in a Tremble', and in a letter to his friend Leigh Hunt, he evoked 'a Whale's back in the Sea of Prose'. But as he walked along the port's medieval walls and looked down the grey waterway, the young poet did not see what Horace Walpole had seen a generation before, 'the Southampton sea, deep blue, glistening with vessels', nor even any of the dolphins which sometimes swam up it. Instead he found muddy shores laid bare at low tide; the sea had run out. 'The Southampton water when I saw it just now was no better than a low water Water which did no more than answer my expectations,' he told his brothers, '– it will have mended its Manners by 3.' Keats's nerves were raw, so he took out his Shakespeare and quoted from The Tempest for solace: 'There's my comfort.' That afternoon he left on the rising tide, sailing to the Isle of Wight where, troubled by the island's strange noises and unable to sleep, he began his long poem, Endymion, filled with moonbeams and snorting whales and leaping dolphins and the tale of Glaucus, the fisherman-turned-god with fins for limbs, whom Endymion frees from Circe the witch.

Keats's contemporary J.M.W. Turner was also stirred by the restless sea. His imagination stained southern skies: he sketched this shore and painted storms off the Isle of Wight, and when he claimed to have had himself lashed to a ship's mast in a blizzard so that he could create a great swirling vortex of waves and cloud – as if he were seeing into the future – he gave the vessel's name as *Ariel*. And in this stormy story, Shakespeare and Turner would in turn influence another writer. Herman Melville's eyes had been damaged by scarlet fever in his childhood and rendered as 'tender as young sparrows'; he was thirty years old, with a career at sea behind him, when he read Shakespeare, discovering a large-print edition of the playwright's works in 1849. As he began to write about his great white whale – his head full of Turner's spumey paintings he'd seen in London that year – Melville read *The Tempest* and drew a box around Prospero's 'quiet words', the magician's wry response to his daughter's naïve exclamation on seeing the aliens:

> *Miranda.*   O! wonder!
> How many goodly creatures are there here!
> How beauteous mankind is! Oh brave new world,
> That has such people in't!
>
> *Prospero.*   ⎡ 'Tis new to thee ⎤

Melville, himself the scion of a colony, saw the prophecy in Shakespeare's drama and in Turner's art: both helped him create the strange, ominous world of *Moby-Dick*. In it, Captain Ahab is a monomaniacal Prospero, and as the opening of *The Tempest* is lit by St Elmo's fire, so the same eerie light garlands his ship, the *Pequod*, in a ghostly glow; animals acquire symbolic meaning – whales and birds accompany the narrative as familiars, swimming and flying alongside the story – and the sea rises up as if with a mind of its own, as it does in Turner's paintings. Meanwhile mortal men pursue their deadly trade: a restive crew sails into the unknown – among them a tattooed cannibal, Queequeg, a kind of Caliban – and Ahab blasphemously baptises his harpooneers in the name of the devil. Nor would the American's fascination with Shakespeare's play end there. At the end of his own last work, Billy Budd, Melville imagines the hanged body of his Handsome Sailor consigned to the deep, entangled in oozy weeds; it is an echo of Jonah's fate in a biblical sea – 'where the eddying depths sucked him ten thousand fathoms down, and the weeds were wrapped about his head' – but also of Alonso in The Tempest, who, believing his son to be drowned, wishes he too was 'mudded in that oozy bed'.

Constantly recreated, constantly re-enacted, The Tempest lived on beyond its creator, passed on from hand to hand. It became a secret cipher, a futuristic shipping forecast, an extended magical spell. It conjured a queer sea out of its strange beasts and its masquerades, and stood against time and tide even as it rose with them in a storm stirred up by a dramatist whose own identity still seems fluid and uncertain.

As a new century loomed, the play gained momentum, gathering clouds rather than diminishing with distance and time. A few years after Melville left Billy Budd on his desk unpublished, another former sailor, Joseph Conrad, drew on The Tempest for his Heart of Darkness, with Kurtz as a terrible Prospero. Two decades later, Eliot embedded fragments of the play in The Waste Land, as if Shakespeare had foreseen the undone. Eliot's work was rivered with the brown god of the Thames and the ship-wrecked, sea-monstered coast of New England he'd sailed as a boy; and as his Madame Sosotris lays out the Tarot card of the drowned Phoenician Sailor – 'Those are pearls that were his eyes' – she warns 'Fear death by water.' Meanwhile, the bones of another sailor, 'a fortnight dead', are picked clean by the creatures of the sea, the slimy things the Ancient Mariner saw down there.

Haunted and haunting, The Tempest accompanied the twentieth century as a parallel rite; few other works of art have been so replicated, remodelled, and re-presented. W.H. Auden reimagined its characters' fates in his verse drama *The Sea and the Mirror*; Aldous Huxley drew on it ironically for his brave new world; and the science-fiction film *Forbidden Planet* turned Ariel into Robbie the robot for an era fearful of its own aliens. And at the end of the darkened nineteen-seventies – as a British satellite named Prospero spun into outer space, following its fellow transmitter, Ariel, launched a decade earlier – Derek Jarman, living in a London warehouse on the Thames and fascinated by John Dee, filmed his alchemical version as 'a chronology of three hundred and fifty years of the play's existence', with Prospero played by the future author of *Whale Nation*, a sibilant blind actor nicknamed Orlando as Caliban, an androgynous man-boy as Ariel, and Elisabeth Welch singing 'Stormy Weather' surrounded by dancing sailors. In this lineage of otherness, filled with hermaphrodites and shape-shifters, it does not seem a coincidence that the director wanted Ariel's songs to be sung by the starman who obsessed me, and who presided over my blue notebook.

The word 'tempest' itself derives from the Latin *tempus*, for time. Everything is new and old on Prospero's island. Like the sea itself. Always changing, always the same.

Out of the blackness obscure noises drift from the docks, booming over the water. The red lights of the power-station chimney blink like an industrial lighthouse, summoning and warning. You can be what you want to be in the dark. For me, it used to be nightclubs under London streets. Now it's another nocturnal performance.

An hour before dawn, before the light starts to stain the summer sky violet, I ride back to the beach. Foxes sidle out of the woods and rabbits flash their white scuts at the approach of my bike light. High in the trees over the shore, a pair of tawny owls converse in screeches. Crows hang in the branches, all angular tails and beaks, as if they'd been born out of the boles. All these creatures own this place in the interregnum of the dark; they should not be anywhere else. No one could have told you when you were young what would happen. They didn't dare. It's enough to realise that what we have lost is still ahead of us. I see things that are not there.

One magical moment; I feel like a penitent. The sea is so still it seems like a sin to break its surface. But I do. Swimming at night, with diminished sense of sight, only makes the act all the more sensual. You feel the water around you; you lose yourself in its sway. Fish bite me, leaving loving grazes.

I turn on my back, watching the stars fall.

I first saw it slumped on the weedy slipway one afternoon. A deer, sprawled at the high water mark. It looked perfect, lying there, thrown up by the tide, staring glassy-eyed to the sky. Had it died trying to swim across from the forest? Or had it slipped and fallen, cloven hoofs clattering on the concrete with panic in its eyes? Perhaps it had been shot, although there was no wound in its russet pelt.

The next day someone had hauled out this sea-deer, this ant-lered seal, and impaled it on the spikes of the metal railings. It hung there by its neck, dangling as a warning, the way farmers nail dead owls, wings outstretched, to barn doors. I wanted to relieve it of this indignity, to take it down from its cross, but I hadn't the strength.

So I waited to see what happened next.

The following day it reappeared on the shore, as if it had climbed down overnight. It was accompanied by a carrion crow, tentatively but intimately pecking away at the flesh, performing the last rites. I wished the bird well, and a good breakfast.

I'd forgotten about the carcase when, a week later, I came across its remains in the surf. By now the body had been reduced to a single strand of vertebrae, picked clean by crabs and gulls. It was down to its essential scaffolding, its skeletal beauty twisted like the ghost of a horned sea serpent lolling in the water. The stubby antlers sprouted from the bulbous, rough-edged rings on the forehead; caught between them was a scrap of fetlock-like fur. Skeins of grey flesh still hung about the skull, scrappily attached to the thin white bone. I had to have it, this grotesque piece of flotsam, something to add to the pile back home, the fragments of blue-and-white china, the clay pipes with the bloom still on them, the shards of misty sea-glass, the chunks of green-glazed medieval pots, the stones pierced with holes.

Using a bit of driftwood to hold down the spine, I pulled at the antlers, twisting and wrestling with them as with a bull. It occurred to me, as I did so, how easy it might be to detach a human head. With a stagger, I succeeded in wrenching off my trophy, my prize for having watched so patiently. I had to gouge out a gelatinous eye before stuffing the skull into a plastic bag and tying it to the back of my bike.

I rode away from the beach, passing walkers who wouldn't guess at my cargo. Back home I opened a hole in the warm brown earth and buried the head up to its antlers. They stood proud of the soil like a pruned rose bush. I piled rocks on top to guard against predators and went back indoors to wait till the antlers sprouted and grew like branches, and as below the surface the skull grew roots which became bones, its lost vertebrae, femurs and ribs all restored, ready to rear up out of the earth, a resurrected, a newly-grown deer of my own.

HEGAZESTOTHESHORE

The runway is spattered with coloured lights, a constellation fallen from the sky. I'm led out into the sharp night air, and take my place beside the pilot. He tells me to slide my seat forward and strap myself in. The dual controls move over my lap, operated by a ghostly co-pilot; incomprehensible dials and LED displays tick and flicker on the console. The plexiglass windows shake with the propellers as we taxi onto the airstrip. We stand ready for takeoff, behind a huge airliner, the kind in which I've just spent six hours getting here. But these last few miles seem the most difficult.

The little plane follows the behemoth, drawing courage from its slipstream. The pilot mutters into his microphone, the runway clears and the wings wobble. Suddenly we are rising over the dark city, made darker by the sea at its edge.

I have to catch hold of my breath, like a child on Christmas morning. I want to turn to the pilot and say, Isn't this *amazing?* But he just stares ahead, wearing his white pressed shirt, quietly suppressing his ecstasy. Everything falls away, all the houses and streets and offices and institutions, leaving only the black water.

The airstrip lights vanish, replaced by winter stars. Orion lurches over the horizon, lazily rising into position, echoing Cape Cod's fragile shape in his starry frame. The night is so clear, made clearer by the cold, that I can see through the Hunter's spaces to the stars he has swallowed, the stars that are being born. We're astronauts for twenty minutes, inside the sky, flying into another system. I look up and down: there's no difference above or below. The sea is full of stars; the stars are full of the sea.

Out of the blackness ahead a line of red lights appears, trembling, beckoning us down. It is a tentative landfall: the only thing below us is sand. We return to earth with a bump. For all I know we might have arrived on another planet. Then the pilot turns round in his seat and says, 'Welcome to Provincetown.'

These past few days the bay has been filled with mergansers. They're saw-beaked, punk-crested birds, forever roving over the sea in their search for food or sex. Just offshore, three males arch their necks in lusty splendour, fighting over a female. Pat says black-backed gulls sometimes take them. Pat is my landlady, although sealady might be a better term. She's lived here for seventy years. She knows this place as well as her own body. It is through her eyes that I see it.

Close up, red-breasted mergansers are even more extreme: big, pugilistic, as though they're cruising for a fight. I see the detached head of one rolling in the tideline. I pick it up, running my index finger along its velociraptor teeth. This winter beach is no place of innocence and play, but a site of carnage and slaughter.

From my deck I hear the forlorn calls of loons drifting across the bay. In the mid-distance is the rocky, guano-spotted breakwater. It was built to protect the harbour, but it was soon colonised by cormorants. They're despised for their droppings that dribble like fishy porridge, and for their supposed depletion of the fishermen's catch; so greedy that they dislocate their jaws to swallow fishes whole. Only Pat sees them for what they are: sentinel creatures she has drawn over and over again, kayaking out to the breakwater and tethering to a lobster buoy, Zeiss binoculars in one hand and a black marker pen in the other.

Pat – who resembles a bird herself, with her shock of silver hair, intense brown eyes and high cheekbones – channels these charismatic spirits. Haughty of our disdain, they pose in portrait after portrait, a cormorant lineage, each profile worthy of a Hapsburg prince. Clamped to the rocks by their claws, heads bent to preen or raised to the sky, they hold out their wings – to cool their bodies as much as to dry their feathers – casting shadows of themselves. Some saw the crucifix in the shape they threw, a symbol of sacrifice; others discerned something darker.

In the opening pages of *Jane Eyre*, published in 1847, Charlotte Brontë's young heroine takes Thomas Bewick's *A History of British Birds* from a library shelf on a winter's afternoon, and hiding in a curtained window seat, loses herself in descriptions of 'the haunts of sea-fowl' in the Northern Ocean, surrounded by 'a sea of billow and spray' and the 'marine phantoms' of wrecked ships.

Bewick's engravings of 'naked, melancholy isles' echo Jane's abandonment as an orphan, 'an uncongenial alien'. Later, when she meets Mr Rochester, she shows him three strange watercolours she has painted. One portrays a woman's body from the waist up, seen through a vapour as the incarnation of the Evening Star; another depicts an iceberg under a muster of the northern lights, overloomed by a veiled and hollow-eyed head. In the third allegory, a cormorant perches on the half-submerged mast of a sinking ship. The bird is 'large and dark, with wings flecked with foam; its beak held a gold bracelet set with gems, that I had touched with as brilliant tints as my palette could yield, and as glittering distinctness as my pencil could impart'. Below it, 'a drowned corpse glanced through the green water; a fair arm was the only limb clearly visible, whence the bracelet had been washed or torn'.

The double-crested cormorant's binomial, for all its Linnaean rigour, is resonant of such gothic airs. *Phalacrocorax auritus* conflates the Greek for bald, *phalakros*, and *korax*, for crow or raven, with *auritus*, the Latin for eared, a reference to the bird's breeding crests. Its common name also reflects the same allusion, if not confusion, as a contraction of *corvus marinus*, sea raven – until the sixteenth century it was believed that the two species were related. Indeed, like ravens, cormorants have a noble antecedence: James I kept a cormorantry on the Thames, overseen by the Keeper of the Royal Cormorants who hooded his charges and tied their necks to stop them swallowing their prey. Bewick called them corvorants and thought their tribe 'possessed of energies not of an ordinary kind; they are of a stern sullen character, with a remarkably penetrating eye and a vigorous body, and their whole deportment carries along with it the appearance of the wary circumspect plunderer, the unrelenting tyrant'; he noted that Milton's Satan perches as a cormorant in Paradise, a banished black angel on the Tree of Life.

The cormorant, whose darkness is implicit in its ability to dive one hundred and fifty feet into the sea, predates any tyrannical monarch; its pterosaur pose evokes the reptilian past of all avians. But to some modern eyes, the cormorant is all too common: a scavenger, a sea crow, or, in the careless calling of the American deep South, the nigger goose. According to Mark Cocker, British anglers call it 'black death', and demand its execution. But all name-calling reflects only on ourselves: we name to know and own, not necessarily to comprehend. We don't even have the right words for ourselves.

Like other animals, cormorants have been forced to share the human stain. Far from eating 'our' fish, the prey they take is of little value to us. Rather, they appear to be attracted to objects that we discard. In 1929, E.H. Forbush, the indefatigable state ornithologist of Massachusetts – a man who acted as a defence attorney for such accused avians (even though he himself ate some of the species he studied) – noted that a cormorantry off Labrador was embellished with objects the birds had salvaged from ship-wrecks, diving like Jane Eyre's bracelet thief to retrieve penknives, pipes, hairpins and ladies' combs. Their finds decorated their nests as if they were making their own artistic comments on our disposable culture.

One autumn morning, after a terrific storm that had swept over the Cape and depressed my spirit with its violence, I woke at dawn to find the sea in front of Pat's house filled with cormorants, hundreds of them. Driven off the breakwater, they'd gathered in a dense raft, avian refugees in an abstract arrangement, each sharp upturned yellow bill, white throat and sinuous neck creating a repeated rhythm, a crazy cormorant expressionism. A scattering of sea crows, marks on the water.

Some perched on the rotting remains of the old wharf, its stanchions reduced, storm by storm, to forty-five-degree angles sticking up out of the water. I watched as the birds rose and fell as one with the swelling waves. Later, I saw them further out. They'd found a source of food, and as the sun glinted on their bobbing bodies they were overflown by herring gulls, a flickering grey layer to their inky black shapes. It was a frenzied, silent scene, watched only by me.

Most mornings, I walk down the beach to meet Dennis and his dog, Dory. Dennis is handsome and everyone loves him. He's sturdy, with salt-and-pepper hair and a trim beard; he reminds me of Melville. When we get off the whalewatch boat it takes us three times as long to get home, because he stops so often to talk to friends and acquaintances in town. Dennis was once a teacher; he did his national service in the coastguard, but has loved birds ever since he was a boy growing up in Pennsylvania. He came to Provincetown by chance, and stayed. Everyone's a washashore here, like the soil itself, brought as ballast to these unstable sands; even the turf came from Ireland, to be laid as lawns for the gracious gardens of the East End.

That morning, as Dennis and I walked towards each other, I saw a bird crouched on the rocky groyne in between us. It had tucked its head into its wings; I presumed it was preening, or sleeping. But as we drew near, Dennis took up his binoculars. Something was wrong. He gestured at me with open hands and then at the cormorant, which slipped off the rocks and into the water.

The bird's bill was lashed to its back by fishing line, and it was tugging pathetically at the monofilament. We followed as it swam parallel to shore. It wanted to return to the land, confused by what had happened to it, as if it might peck off its trusses. But each time we approached, it went back to the water. Dennis was not optimistic. 'It'll just keep swimming out – or it'll dive,' he said.

I waded into the sea. Dennis ran further up the beach, staying close to the bulkheads to keep his profile low. I tried to splash the cormorant ashore. It worked: the bird made for the beach and Dennis dashed towards it, unafraid of its flapping bulk.

Suddenly there it was, in our hands. A startling sapphire circle around a green cabouchon eye; a fractured sharpness, staring back unblinking. Up close, every feature took on the definition that Pat had drawn: the yellow-tipped bill and its hooked tip, the matt black wings. Primeval enough from afar, this near the bird looked even more like an archaeopteryx on the beach; evolution in our fingers.

Any bird exists apart from us: unmammalian, and therefore uncanny. Yet I could imagine myself a cormorant mate, entranced by this handsome fellow, building a nest on the rocks, proudly holding our bills in the air in celebration of our cormorantness. We took the bird to the deck of a beach house under construction, where a workman produced a knife. Swiftly, Dennis cut the line and pulled the hook from the cormorant's mouth. Blood trickled out, bright and fresh against the black feathers. Dennis was promptly pecked on the thumb for his trouble, drawing his own blood in turn. I unwound the bird's bound wings. In a second it was free, half running, half flying back to the water for its lunch.

After a day of louring greyness, when the town seems bowed down by the low pressure – 'Everyone I meet says they have to go home and sleep,' says Pat – I retreat to my studio in the timber house. Its double gables rise over the beach like some Nordic chapel or a barn raised by settlers, held up by a pair of cinderblock chimneys. I sleep in its eaves, in an attic like a chandler's loft or a ship's prow. At night I climb up to my platform bed via a wooden ladder, ascending to my dreams, descending in the morning, clambering down backwards the way I would at the stern of a boat.

The house is a fragile and sturdy construction, built to withstand three hundred and sixty degrees and three hundred and sixty-five days of weather. In the winter the wind worries at its windows, with their layers of glass and screens, grooves and latches, a complicated, ultimately ineffectual system of defence. No one wins against the wind, not even these wind's eyes. Running in front of the house is the deck, a wide wooden stage over which Pat lays a path of threadbare yard-sale rugs to stop its splinters from entering her bare feet. They lend the boards a tattered luxury, like some trampled boudoir. Stirred by the wind and rain, they take on a life of their own, rucking like a ploughed field out of which ever-larger splinters sprout as spiky seedlings.

The whole house is still partly tree. The knots in its walls have fallen out over the years, leaving spy-holes and escape routes for whatever creeps and scratches about inside. There are so many compartments, cupboards, stairs and crawl spaces – so many spaces within spaces – that there could be colonies of creatures living under its eaves. Even as I write this, I discover a narrow staircase which I had never seen before in all the years I've been staying here, hidden in a cupboard and leading to the top floor like some secret escape route. And when I open the built-in linen closet on the first floor, a cat hisses at me, leaps up a shaft and disappears into an interior where, for all I know, an entire community of feral felines might reside. I sleep with bare boards next to my head, stamped with the timber merchant's marks,

MILL 50 MILL 50 MILL

W.C.

L.B. ®

UTIL 3/4 W.R. CEDAR

and occasionally silverfish run up and down the western red cedar, their filigree antennae feeling their way like tiny lobsters, while mice scratch in the eaves. I feel the weather and the sea through the wooden walls and the way the day arrives and the night leaves, and there's sand instead of biscuit crumbs in my sheets. Sometimes the whole house becomes a woodwind instrument played by a demented child. Doors rattle, urgent spirits seeking admission. Timbers creak as a ship caught in ice; articulated chimney cowls squeak like weather vanes turning in the wind. The house reverberates as though remembering how it was built, an echo chamber resonant with everything happening outside and everything that ever happened within. It may be inanimate, but it makes me more alive, this big beach hut. How could anyone not feel that way, knowing that out there is the sea, and all land is lost to the horizon?

The front hits us, head on. The waves, which yesterday lapped the footings of the house, turn over themselves in their remorseless assault on the bulkhead that acts as a buffer between the house and the sea. Town regulations, designed to allow the shifting sand its sway, mean that even the most luxurious decks and dining rooms are temporary arrangements. Pat's house, now in its sixth decade, was built to be part of rather than apart from the water; in stormy spring tides the sea actually runs right underneath it, disdaining its foundations. By the end of the century all those exclusive properties and ramshackle shacks alike will yield to the waves. 'The truth is,' the philosopher Henry David Thoreau wrote on one of his visits to the Cape, 'their houses are floating ones, and their home is on the ocean.'

Directly in front of the house is a raft tethered by a chain to the sea bed. It's another stage, a four-foot-square island of performance. In the winter seals lounge on it, their doggy heads and flappy feet held in the air to keep warm. Summer visitors think the raft is built for human swimmers; they soon realise that it's covered in deposits from its other tenants, the eider ducks that take up winter leases, and for whom it is a safe perch even when it rocks wildly in high seas.

Pat and I watch a duck and drake circling the float as if sizing it up. The male makes the first move, followed by his partner. They stake out their separate corners, like a couple seeking their own space. Another male appears with his mate; she is allowed on board, he is rebuffed by the first male. It's a stand-off. There follows a ritual puffing up of chests and fluttering of wings, like a contest on the dance floor. The inevitable compromise is reached, and the newcomers are admitted. Soon, in the niceties of eider choreography, a third couple arrive and the same rite is observed. All their gestures and cooing, which seem quaint to our anthropomorphic eyes – as if they were saying to each newcomer, 'No room, no room' – are in fact grim and determined expressions of potential violence and struggle for precedence.

Eiders are another of this shore's animal spirits. They preside, like the cormorants and the seals, imbued with their own inscrutability. The raft is their portal: I imagine them diving off it and coming up in a willow-pattern world to reassume their imperial presence, shrugging their lordly wings as they do so. They may be the largest of the ducks, but they're also the fastest bird in level flight, able to fly at seventy miles an hour against fierce nor'easterlies. They are endlessly interesting to me, seen from my deck or through my binoculars. Their heads slope down to wedge-shaped bills, redolent of Roman noses or a grey seal's snout. Their black eye patches and pistachio-green napes look like exotic make-up, although Gavin Maxwell thought that they wore the full-dress uniform of a Ruritanian admiral. Their table manners are hardly refined: they use their gizzards, lined with stones, to grind and crush the mussels and crabs which they swallow whole. Birds as machines.

They too have suffered. In Britain they were used for target practice during the Second World War; thousands of them, lying in rafts on the sea, were blasted away. On the Cape that winter I find many eider carcases strewn across the sand, ripped open and spatchcocked, as if the violent cold were too much even for them, despite their downy insulation. One victim's eyes have long since puckered into blindness, but its nape is still tight, like the back of a rabbit's neck, more fur than plumage. Eiders are still harvested for their air-filled feathers to make quilts and coats, 'robbing the nests and breasts of birds to prepare this shelter with a shelter', as Thoreau wrote. They tolerate our appropriation; they have no choice. But while we may have our uses for them, their features speak of something unknowable.

Perhaps it is those eyes. Yes, it's those eyes. It always is. They take in the whole of the world, even as they ignore it.

Held out into the Atlantic, Cape Cod is a tensed bow, curled up and back on itself, a sandy curlicue which looks far too fragile to withstand what the ocean has to throw at it. Battered by successive storms, its tip has been shaped and reshaped for centuries. It is only halfway here, and not really there at all. It is porous. The sea seeps into it.

This is where America runs out. Sometimes, if the light is right, as it is this morning, the land across the bay fizzles into a mirage, a Fata Morgana stretching distant beaches into seeming cliffs, floating dreamily on the horizon. The further away you are, the less real everything else becomes. This place takes little account of what happens on the mainland; or rather, puts it all into perspective. It is a seismograph in the American ocean, sensing the rest of the world. Not for nothing did Marconi send out his radio signals from this shore; he also believed that in turn his transmitters might pick up the cries of sailors long since drowned in the Atlantic.

The inner bay arches around from the lower Cape, losing people as it goes. From empty-looking lanes where signs politely protest THICKLY SETTLED, as if there might be as many inhabitants as trees, you pass through Wellfleet's woods and second homes to North Truro's desultory holiday cottages on the open highway, as lonely as Edward Hopper's paintings, and on to Provincetown, where the land widens briefly before dwindling to Long Point, a spit of sand as slender and elegant as the tail on the tiny green spelter monkey that sits above Pat's woodstove. Long Point Light stands on the tip, a square stubby tower topped with a black crown lantern – it might welcome or warn off visitors, it doesn't really matter which. Once you're here, you never leave. This is the end and beginning of things.

I first came to Provincetown in the summer of 2001. Invited here by John Waters, I was in town for just five days; I had no idea then what they would mean to me now. Like some perverse mentor, John initiated me into the secrets of the place. We drank at the A-House, where grown men groomed one another's bodies like animals eating each other's fleas; and we drank at the Old Colony, a wooden cave that lurched as if it was drunken itself; and we drank at the Vets bar, where the straight men of the town took their last stand in the dingy light. On hot afternoons we hitched to Longnook, using a battered cardboard sign with our destination scrawled on it with a Sharpie, waiting on Route 6 for a ride. Once a police car stopped for us. We sat on the caged back seat like criminals and when we arrived, John said, 'We've been paroled to the beach.' He looked out over the ocean and declared it to be so beautiful that it was a joke. When he rode down Commercial Street on his bike with its wicker basket like the Wicked Witch of the West, I heard someone call 'Your Majesty' as we passed.

It was only at the end of my stay, about to take the ferry back to Boston, that I decided to go on a whalewatch. I stepped off the land and onto the boat. Forty minutes later, out on Stellwagen Bank, a humpback breached in front of me. It still hangs there.

It is not easy to get here. It never was. For most of its human history, Provincetown was accessible only by boat, or by a narrow strip of sand that connected it to the rest of the peninsula. And even when you did arrive, it was difficult to know what was here and what was there; what was land and what was sea. Maps from the eighteen-thirties show a place marooned by water, its margins partly inundated. There was no road till the twentieth century;

the railway once raced visitors to Provincetown, but that was abandoned long ago, as were the steamers that brought trippers from New York. Nowadays ferries run only from May to October, and the little plane can be grounded by lightning striking the airstrip or fog shrouding the Cape. Provincetown is where Route 6 starts, running coast to coast for three and a half thousand miles all the way to Long Beach, California. But it was renumbered in 1964, and now North America's longest highway seems to peter out in the sand, as if it had given up before it began. It is a long, long drive here from Boston, and the road becomes progressively narrower the further you go, curling back on itself till the sea presses in from all sides, leaving little space for tarmac, houses, or people. No one arrives here accidentally, unless they do. It is not on the way to anywhere else, except to the sea.

The lost people who find their way here discover the comfort of the tides, anchoring endless days which would spin out of control, faced with the wilderness all around. My time is defined by the sea, just as it is at home. But instead of having to cycle to it, I only have to roll off my bed, and stumble down the wooden steps. I sniff the air like a dog, and lower myself off the bulkhead. The eiders coo like camp comedians. The water is the water. I turn on my back, face up to the sky, the monument high on the hill behind me marking the arrival of the Pilgrims who set sail from Southampton for this shore three centuries ago. I swing my body towards it like a compass needle. It's as though I'd swum all the way here. I count my strokes. The cold soon forces me out. I climb back upstairs to boil water for tea, holding my hands over the glowing electric element to restore the circulation enough to let me write.

On my desk sit the objects that spend my absence stashed away under the eaves like Christmas decorations. A swirly green glass whale I bought from the general store. A nineteen-twenties edition of *Moby-Dick*, a faded coloured plate stuck on its cover. A slat of driftwood found on the beach, with layers of green and white paint peeling away in waves. A tide table pinned to the wall, although I don't really need it. My body is tuned to the ebb and flow; I hear it subconsciously in my sleep, and feel it wherever I am in town. Everyone else feels it too, even if they think they don't. It stirs me from my bed and summons me to the sea, whatever the time of day or night.

I've spent many summers here; winters, too. I've seen it out of season, when the people fall away with the leaves to reveal its bones: the shingled houses and white lanes lined with crushed clam shells as if they led out of or under the sea. Squeezed on all sides by the sea, houses here are built efficiently, like ships; in a place like this, you don't waste space or resources. An artist's studio has drawers built into the risers of the stairs, turning them into one big ascending storage unit. At another cottage, over a glass of gin, I admire a galley kitchen with plates stored on sliding racks. The artist tells me they were designed by the previous owner, Mark Rothko. 'He made us promise never to change them.'

Provincetown may be a resort to some, but it is at its best at its most austere, when everything is grey and white and hollow, and you can peer over picket fences into other lives; backyards full of buoys or old trucks where a century ago there would have been nets and harpoons. Once this was an industrial site – hunting whales, catching fish. Then it emptied, forgotten by the future which left its people behind, the insular people Melville knew, 'not acknowledging the common continent of men, but each Isolato living on a separate continent of his own'.

On warm summer nights, Commercial Street, one of only two thoroughfares that thread through the town, is an open, sensual place; in the winter, when the cold comes inside and won't leave of its own accord for half the year, the rawness returns it to a dark lane, winding nowhere. In 1943, when the town was further shadowed by the threat of air raids and German landings – as if its held-outness was a kind of sacrifice to the war going on across the ocean – the young Norman Mailer walked down the blacked-out street and back into the eighteenth century, or at least what he felt was 'a close intimation of what it might have been like to live in New England then'. It's difficult to imagine an inhabited place so empty. Even during the day in the twenty-first century, a chill sea mist can envelop its springtime streets – all the seasons are delayed here – filling the glowing white lanes with ghosts. There are spirits throughout this creaking old town. You see their shadows on stairs, shapes out of the corner of your eye. In the winter, they walk down the street. They're there in the summer too; they just look like everybody else.

The sea accelerates and stalls time. This town has altered in many ways, even in the fifteen years I have been coming back to it, for all that it stays the same. I'm never quite sure when I return that I will be accepted by its people, its weather, its animals, or that anyone will remember me, and am always surprised when they do. I'm always arriving and always leaving; as my friend Mary across the street says, the moment you arrive anywhere is the start of your departure. Life here is measured by the waiting for spring, the longing of the fall, the waiting for summer, the longing of winter; everything is restless, like the sea. Sometimes it seems so perfect that I wonder if it even exists, if it isn't all a vision which rises through the plane's windscreen as I arrive and disappears off the ferry's stern as I leave; and sometimes I wonder why I come here at all, when the wind whines and voices bicker, when cabin fever takes over and doors blow back in your face.

This is not a kind place. It leaves its inhabitants biopsied, like the scars in skin too long exposed to the sun. Lungs collapse with too much cold air. Like their forebears, they suffer for presuming to live on this frontier. It is a continual challenge to body and mind. A place of dark and light, day and night, storms and tides and stars; a place where you have to feel alive, because it so clearly shows you the alternative.

Pat's house is so much of a boat that it might have been floated across from Long Point, as houses were in the nineteenth century, or been trawled out of the bay, like the whaling captains' mansions down in New Bedford, 'brave houses and flowery gardens, that came from the Atlantic, Pacific, and Indian Oceans, harpooned and dragged up hither from the bottom of the sea'. Inside her studio, Pat's state-of-the-art kayak is slung from the rafters alongside an older, wooden model, both hanging there like stuffed crocodiles in a cabinet of curiosities. A large plastic sheet is stretched between them to catch the rainwater that drips from the roof. With typical ingenuity, Pat has rigged up an intricate series of lines and pulleys, along with a plastic tube draining the swelling whale belly of the sheeting like a catheter into a hanging bucket which, when full – as it is from last night's storm – can be lowered to be emptied, just as Pat's kayaks can be lowered, ready for the days when she would paddle out to the Point and beyond, not really caring about coming back.

The rigging turns her studio into an inside-out yacht. It is a kinetic work of art in itself. Lightbulbs dangle from electric cords like the lures of angler fish, but there are no bright lights inside because all the light is outside. Doors slide to reveal store cupboards capable of stacking huge canvases like theatre scenery. The whole house is slotted together, a serious plaything, a place to work and be and think and drift along with the seasons. It is part of her body, an extension of her self. It is entirely practical, fitted out rather than built. On the studio walls hang Pat's paintings of the view outside: the same scene painted again and again, like the cormorants; the same proportion of sea and sky, the same dimensions divided between air and water, in mist and fog and snow and moonlight. They are not so much paintings as meditations. They look through the moment of seeing – the falling fog, the drifting snow, the rising moon. They are the sea reduced to its essence. They are not concepts. Pat's husband Nanno de Groot told her, 'Analyse your stupidity.' 'I don't think about anything else when I work,' Pat tells me. That's because her work is not like anything else.

She uses no brushes, but applies the paint with a knife; removing, rather than adding, to reveal what was there all along. The paint is flattened, smoothed, pushed in; you can feel the power of her hand and arm and shoulder behind it. But at the same time the colour – the medium between what she sees and what she puts down – rises rhythmically like the waves and clouds it re-presents, grey and green and white and blue. Pat paints the memory of the actuality of the thing – the thing that lies out there. It all comes down to the water. When I admire one painting of a dark sky and a silver sea, she says, 'I waited half my life to be able to paint that.'

Everything is here; everything disappears. Every window is a frame for her work: windows in her dining room, the windows she looks through from her bed, the windows in her bathroom, the windows in her head. They all admit possibilities and impossibilities; work-in-progress. Her mind is laid out here. You can follow the trail of her imagination from her studio and into her house. Half-squeezed tubes of paint lie under Buddhist prayer flags, next to scraps of sun-yellowed paper and rolls of masking tape, tiny palette knives and piles of fading *National Geographic* magazines. On a work table is a clam shell in which a finch is curled, quietly sleeping, all but breathing, its perfect feathers still blushed pink.

Pat is in her eighties now. She doesn't paint much any more. She doesn't have to. When she talks to me in the morning, the sun already turning the deck hot by eight o'clock, she carelessly raises her leg above her head in a yoga pose. She weighs one hundred pounds. She is wired as much as muscled. She still sunbathes naked in the dunes, where national park rangers have threatened to issue her with a ticket for flouting the bylaws. Pat tells them they can do what they like; she's been doing this for seventy years, and she's not about to stop now. She walks barefoot all day – 'Bare feet are older than shoes,' as Thoreau says – padding along the beach, more animal than human. All the time I've known her, she has always kept German Shepherds close to her. They are wolves in disguise, just as she is half dog herself. It's taken me fifteen years to hear her story; she keeps it in reserve, hidden in her cupboards and drawers. The withholding only makes the past more present.

Pat was born in London in 1930, but in 1940, when she was ten years old, she and her brother were sent to America by their parents. She still finds this extraordinary, as if she can't quite believe it even now. Her father, Ernald Wilbraham Arthur Richardson, was born into the landed gentry in 1900; his own father, who had served in the South African war, was English-Welsh, and his mother was Irish; the family had a large country estate in Carmarthenshire. Ernald followed the progress of his class, from public school to Oxford, but his passion was skiing, and he was an Alpine skiing pioneer in the nineteen-twenties, photographed on the slopes as part of the British ski team, a dashing young man. In 1929 he had travelled to the US, where he met and married Evelyn Straus Weil, a smart, chic young New Yorker of twenty-three with dark hair and big bright eyes whom her daughter would describe as a flapper. She had a decidedly more cosmopolitan background than her English husband.

Evelyn's grandfather was Isidor Straus, a German-born Jew who had joined his father, Lazarus, in New York in 1854. There the family forged a remarkable partnership. Lazarus Straus went into business with a Quaker from a celebrated Nantucket whaling family, Rowland Hussey Macy. Their department store boomed. In 1895, Isidor and his brother Nathan took over ownership of the store. They had now become a firm part of American life. Both were philanthropists; Isidor had raised thousands of dollars to aid Jews threatened by pogroms in Russia, and Nathan's son, also called Nathan, would try to get visas for Anne Frank's family. Isidor, Pat's great-grandfather, became a member of Congress and turned down the office of Postmaster General when offered it by President Grover Cleveland. Isidor was devoted to his wife, Ida, and their seven children, among them Minnie, Pat's grandmother.

On 10 April 1912, after a winter spent in Europe, Isidor and Ida boarded a new luxury liner at Southampton, bound for New York. Five days later, in the early hours of 15 April, as *Titanic* struck an iceberg and began to sink 375 miles south of Newfoundland, the couple's devotion to each other became a modern legend. Ida declined to get into a lifeboat without Isidor. And since there were still women and children on board, Isidor refused the offer of a place in a boat alongside his wife.

'I will not go before the other men,' he is reported to have said, in formal, polite insistence. 'I do not wish any distinction in my favour which is not granted others.'

Ida sent her English maid, Ellen Bird, to lifeboat number eight. She gave Ellen her fur coat, saying she would not need it herself: 'I will not be separated from my husband. As we have lived, so will we die, together.'

The couple went and sat on a pair of deckchairs. It was, according to those who witnessed it, 'a most remarkable exhibition of love and devotion'. I see that determination in Ida's face and Pat's: the same brow, the same eyes.

Isidor and Ida, along with fifteen hundred other souls, perished in a sea described as a white plain of ice. Most died of cardiac arrest after a few minutes in the minus two degrees water. One rescue ship came across more than one hundred bodies in the fog, so close together that their lifebelts, rising and falling with the waves, made them look like a flock of seagulls bobbing there. Isidor's

body was recovered and brought back to New York; his funeral was delayed in the hope that Ida's body might be found. It never was: fewer than one in five were, and of those, only the corpses of the first-class passengers were worth bringing back, since their relatives could pay. The rest were tipped back into the sea.

Nearly thirty years later, Pat's mother Evelyn – known as Evie – sent her and her brother across that same ocean – itself a dangerous journey during wartime; in June 1940 the ship they sailed on, SS *Washington*, had been stopped by a German submarine on an earlier voyage taking back Americans who had been warned to return to the US without delay or stay in Britain at their own risk. (As a Jew, Evie would have been concerned at what might happen if the Germans invaded. Ten years later, the same ship would sail from Southampton to New York, carrying survivors of the Holocaust.) The liner's deluxe interior – its staterooms, ballroom and library – was filled with families. Archive film shows the deck piled high with trunks and suitcases, and children being led off the ship on arrival in New York with teddy bears in their hands or in prams and pushchairs. Their evacuation was done for their safety, but Pat came to believe that both her mother and her father wanted to conduct their various affairs unencumbered by their offspring. It had not been a happy marriage. Her parents had divorced in 1936, leaving Evie to conduct an affair with Ralph Murnham (later the queen's surgeon) before marrying her second husband, Sebastian de Meir, son of a Mexican diplomat, in 1939; he enlisted in the RAF and died when his bomber plane was shot down over the Netherlands in 1942. Evie, who had taken up nursing in London during the war, moved back to New York in 1943.

Pat had always felt abandoned. 'I was a refugee,' she says. As a girl, growing up in St John's Wood in London, she had hidden in the park, imagining herself as an animal; one of the first books she remembers reading, in the nineteen-thirties, was about a boy who was shipwrecked and stranded on an island where he was brought up by wolves. She wanted to be that boy. Her parents did not care about animals; nor did her nanny, whom Pat remembered wearing a sealskin coat. Pat's mother must have been beautiful and chic. She gave Pat a beaver collar, but Pat refused to wear it, and wouldn't even touch her mother when she wore her fur coats. Pat remembers when Evie showed her a rug made of cat fur. 'She knew I loved cats. She hated them.'

A faded photograph in Pat's bedroom shows 'Captain E.W.A. Richardson, February 1944', now serving in the Queen's Regiment, dressed for the Canadian winter in a white wool duffelcoat as thick as snow. His face is broad and handsome and British. He glows.

Evie's life was as unstable as the times. In 1945 she married Martin Arostegui, a Cuban publisher whose previous wife, Cathleen Vanderbilt, an alcoholic heiress, had died the year before. Within a year they had separated, and Evie married George Backer, an influential Democrat, writer and publisher of the New York Post. Like his friend Nathan Straus, Backer had worked to save his fellow Jews: in 1933 he had travelled to Poland and Germany to help Jewish refugees flee the growing Nazi menace, and he was awarded the Légion d'Honneur by the French government in 1937 for his efforts. 'It is horrible to think,' he would later recall, 'how responsible we were for all that happened. The ships were there and the people were not saved.'

But Evie's world was Manhattan, a world of money and powerful people. Her husband's friends included William Paley, the head of CBS, and Pat recalls that another friend, Averell Harriman, US ambassador to Britain and heir to the largest fortune in America, had also attempted to seduce her mother. Described by the *New York Times* as 'a small, fast-moving woman … amusing, gay and sharp-tongued', Evie drew on her sense of style and her impeccable contacts to become an interior decorator; her clients included Kitty Carlisle Hart, Swifty Lazar and Truman Capote. The pictures in her apartment on the Upper East Side, at 32 East 64th Street, were hung low and small-scale furnishings were chosen to reflect Evie's five-foot-four stature; she moulded her environment to her requirements, just as her daughter would do. Capote named her 'Tiny Malice' for her quick wit. She created a lavish, almost visceral apartment for the writer on the UN Plaza, painting the drawing room blood-red and installing a Victorian carved rosewood sofa, a $500 Tiffany lamp, and a zoo of mimetic and dead animals, from a bronze giraffe and china cats to jaguar-skin pillows and a leopard-skin rug. I can hear Pat's horror. Cecil Beaton called it 'expensive without looking more than ordinary'. But Capote approved, and asked Evie to design his Black and White Ball, the most famous, or notorious, party of the twentieth century, notable for the fact that, despite Evie's recommendation, Capote declined to send an invitation to the President.

She and Capote were snapped arriving at Manhattan's fashionable Colony restaurant. Truman wears a bow tie and horn-rimmed spectacles. He greets the paparazzi, his notorious guest list in his hand; how the magazine editors longed to see that roster. Evie is by his side, thin and chic, conspiratorial in dark glasses. They're both diminutive, yet the centre of all attention. They retreat to one of the coveted back tables – the Cushing sisters on one side, James Stewart on the other – to plot the party. Margaret, Duchess of Argyll is added to the list – Evie says it never hurts to invite a few duchesses. Later, Capote crosses her off too.

The venue was the Grand Ballroom at the Plaza Hotel, celebrated in the twenties by F. Scott Fitzgerald. The event exceeded any of Gatsby's parties. Evie ordered red tablecloths and gold candelabra entwined with 'miles of smilax', a green vine. The guests wore masks, barely disguising their celebrity: Lauren Bacall and Andy Warhol, Frank Sinatra and Mia Farrow, Norman Mailer and Cecil Beaton, Henry Fonda and Tallulah Bankhead. There were Guinnesses, Kennedys, Rockefellers and Vanderbilts, and it was a marvellous party; its ghosts might still be dancing now.

Evie was never more in her element; her daughter couldn't have cared less. High society was far from how Pat wanted to live; now she looks at those photographs, those thin society queens, with disdain. She was, and still is, a teenage rebel, a dropout, and had been ever since she first came to Provincetown, at the age of sixteen. In 1946, her mother had rented John Dos Passos's house in Provincetown's East End for a year, having been alerted to the Cape's allure by Dorothy Paley, wife of William Paley and friend of Dos Passos. It was a heady introduction. It changed the course of Pat's life.

I find it almost impossible – but not quite – to imagine what this place was like then. Its lanes seemed part of the country; many still do. Fishing and whaling had left the remote town open to other influences; a wilderness which allowed the wildness of its inhabitants. Pat worked in the bookshop, but was fired because all she did was read. Then she worked as a waitress in the Flag Ship, where the bar was a boat, and where the owners didn't feed her. Her mother complained that Pat was losing weight – less attractive to the rich Jewish boys with whom she tried to pair off her daughter. Pat would rather go out on Charlie Mayo's boat and sit on the fly bridge, watching the whales and birds. Charlie lived across the street. He was a champion fisherman; his family, part Portuguese, had been on the Cape since 1650. His father had hunted whales, as did Charlie; he only stopped when he harpooned a female pilot whale and heard the cries of her calf beneath his boat. Pat saw Charlie as her surrogate father. They talked and fished. Her mother disapproved; she thought Mayo was a communist. Pat didn't care. She cared about the sea.

Evie had sent her to Austria, in the way that young women of wealthy families were sent to finishing school. Vienna in 1948 wasn't a good choice for a girl like her; there were no zithers playing, and a former Nazi officer tried to rape her when he discovered she was Jewish. Pat came back to college at Pembroke, outside Boston. She loved riding and skiing. But her mother took her away, and her stepfather arranged for her to go to the University of Pennsylvania in Philadelphia, studying English literature and journalism. Pat felt abandoned all over again.

After graduating in 1953, Pat went to spend time in Benson, Arizona, close to the Mexican border, working on a ranch with the horses she loved. 'I was outside all the time I wasn't sleeping.' She planned to go to Taos, where Georgia O'Keeffe had worked; Pat had an artist friend there, and thought that she might learn to paint. But her mother protested about that, too, and Pat was persuaded to go to Paris, where she worked for the *Paris Review* and George Plimpton, typing up Samuel Beckett's manuscripts, riding round the city on a bicycle. She lived in a tiny room at the Hôtel Le Louisiane in Saint-Germain, where Sartre stayed and where the sight of her fellow tenant Lucian Freud, a man who had the look of a raptor, scared her. 'I was not very hip and was hideously shy.' On an assignment to Dublin, where her father now lived, Brendan Behan hit on her in a bar.

No wonder. She was a fine, fierce, uncaptured muse, waiting for the moment. In New York she worked for Farrar, Straus & Giroux; Roger Straus was her cousin. She lived in a walk-up at 57 Spring Street, north of Little Italy, which was pretty funky and a long way from the UN Plaza; the building still stands, hung with its fire escape, two doors down from a restaurant called Gatsby's. The rent was twenty-five dollars a month. Pat would fight with Italians for parking space for her black business coupé, and thought the poor Puerto Rican families were happier than her. On Friday nights she'd leave the office and drive all the way to Mount Washington to ski.

When she had to leave her apartment she moved to the Chelsea Hotel, setting up an office in her room. She took a course in book design at New York University with the designer Marshall Lee. 'He was a good teacher.' It was the only formal training she had. She excelled at it. Even now she'll hand me a new book from her packed shelves and flick through it, expertly analysing its qualities. Her designs were simple and smart. For Thom Gunn's collected verse she created a helical motif, a graphic contrast to the poet's photograph on the back, showing the bearded Gunn crouching in a field, shirtless, in tight jeans, a leather belt loaded like his name. Bennett Cerf, the celebrated founder of Random House, told her mother how brilliant Pat's designs were. Evie just asked her daughter, 'Exactly what is it that you do?'

Pat and Evie. The pearls. The champagne. The lighted cigarette.

Manhattan could never rival Provincetown, and Pat kept coming back. In the summer of 1956 she met Nanno de Groot, a Dutch-born artist, for the second time, having met him briefly when she was eighteen and he was living with his third wife, Elise Asher, in the West End, next to friends of Pat's. That second meeting was memorable: 'When I woke up he was sitting on top of a weir pole, on his feet like a bird looking out to sea, waiting for me.'

He was an imposing figure, forty-three, six foot four, often bare-chested, and always bare-footed, as Pat would be. He'd been to nautical school in Amsterdam and had served on submarines, but was now the artist he had always wanted to be, part of the New York circle of de Kooning, Pollock, Franz Kline and Rothko. 'We spent that week together,' says Pat. She moved into his farmhouse in Little York, New Jersey. They got married on Long Island two years later; the reception was held at the Backers' summer house on the sea-surrounded Sands Point – Daisy Buchanan's East Egg. In the winter they lived in Little York; Nanno painted, Pat went to work in the city. In the summer, they'd return to Provincetown, living in a three-room cabin in a field at the end of town. A photograph shows them there: white light, Nanno naked to the waist, Pat svelte and tanned too, feet up on a table. Nanno painted the trees and the land and the sea – the passing seasons, fishermen's nets drying in the fields – in between working as a mate on Charlie's boat.

It must have been mad and idyllic and frustrating and ecstatic, this life together, in the dunes, on the streets, at sea. Pat remembers 1961, the summer with no wind, when they'd go out on the boat in the glassy calm, so clear you might reach down and pluck fish out of the depths. It was 'a visual onslaught', Pat says in a later, filmed interview in which her style emerges, a mix of bohemian smartness and concentrated beauty. With her wavy, centre-parted hair she might be one of the Velvet Underground, or a Renaissance model. She looks straight at the camera, but sees something else in the distance. She talks about Nanno, who wrote, 'In moments of clarity I can sustain the idea that everything on earth is nature, including that which springs forth from a man's mind, and hand.' He read Robert Graves's *The White Goddess* and painted birds; birds which, as Pat says, 'he felt he might have become', just as she might have become a wolf.

From one of her studio shelves, low down where the cats prowl, Pat pulls a brown envelope, and from it a photograph of herself and Charlie.

It's 1961. There's no wind. The five-hundred-pound tuna dangles between them, suspended by a rope around its tail, so huge and bug-eyed, so stuck over and spiny it's hard to believe it's not cut out and glued on. They each hold a fin, these two anglers, smiling for the camera, proud of their catch.

Pat has a huge rod and reel. As slight and chic as she seems, in her rolled-up jeans, checked shirt and suntan, it was Pat, not Charlie, who did all the work; Pat who struggled to hoick the bluefin out of the sea and onto the deck; Pat who was given the trophy by the state governor for her prize catch, seen in another press photograph, dressed in a dark silk shirt-waister, as shiny as a fish, her glossy hair in curls. She looks like Hepburn or Bacall, gamine and self-assured, with Charlie as her Bogart.

It was Nanno and Charlie and Pat, out fishing, part of the sea. In 1962, Nanno and Pat built this big house, created to enable thin slivers of art. They bought the land for six thousand dollars. Pat drew up the plans and the house grew up from the shore. It didn't so much look out to the sea as the sea looked into it.

'It wasn't conceptual,' Pat says. 'It rose up out of the mud.' Locals thought it was impractical. It seemed built out of belief alone. A factory of the imagination.

That same year came Nanno's diagnosis, 'and everything that goes with that'. Photographs show him bundled up to the neck, sitting on the deck, while the house rises pristine behind him, full of light and space. Living with lung cancer, he painted his last painting, of the sea, the large canvas laid flat, supported on stools. It showed the harbour flats drained at low tide. For the first time, he painted no horizon.

'It was,' said Pat, 'his last word on the subject of painting.' They moved into the house at Thanksgiving, 1962. They were there together barely a year. The following Thanksgiving – just days after President Kennedy was shot – there was a terrible storm which worked its havoc through three high tides. 'It took the bulkhead, the deck, and almost undermined the house,' Pat recalls. A month later, that Christmas, Nanno died.

Pat had his coffin constructed from red cedar left over from the building of their house; as if he were being launched out to sea, like Ishmael. Nanno's tempestuous scenes of the Atlantic shores still hang on these walls: Ballston Beach bursts with energy, as if it were just a window on the wall looking over to the ocean side of the Cape. Every cupboard, every drawer, every eave of this house is filled with art. Art seeps out through the knots in the wood, like the sea under the floorboards.

There were parties here back in the sixties and seventies, recorded in flaring home movies and remembered in the stories of those who attended them and spent a night in gaol for disturbing the peace. There were psychedelic drugs, and when Pat invited jazz musicians, like her lover, Elvin Jones, she'd find rotting fish on her doorstep, left by folk who took offence at her having brought black people to town. Nina Simone visited; I imagine coming downstairs and finding her sipping tea at Pat's long table, talking in her rich voice. A faded photograph pinned to the wall shows Pat and her friends playing congas out on the deck. The drums still stand in her living room, but they haven't been played in a while.

Pat had other visitors to attend to. In 1983, a lone orca appeared in the bay. It was a female, apparently habituated to humans; some thought it was an escapee from a military marine mammal programme, a dolphin draft-dodger. It was the biggest animal she would meet. Pat would kayak out to meet it and drew it over and again, this time using her black marker on flat stones. With the fin rising next to her boat, Pat held out a flounder to her friend.

Others were less considerate when the whale came in close to the pier. 'Someone poured bourbon in her blowhole,' Pat says. After that, the harbourmaster drove the whale back out to sea.

This house is rebuilt with every season, growing layer upon layer. Giant jade and ficus plants tower in the interior, tended by rainwater collected from the roof. Buddha sits in his lotus position in the garden. The outside comes inside. In the yard, self-seeded trees shade the graves of departed dogs; great strings of blue lights illuminate their branches as night falls. Robins and cardinals take refuge up there from the cats to whom this house really belongs, familiars to their mistress.

It is the very antithesis of the order her mother created in fashionable Manhattan. Books and catalogues rise in piles on every step of the stairs. Dusty drawers are filled with cormorants cawing and clamouring to get out. If Pat no longer paints, perhaps it is because she has said what she needed to say. Now she collects stones from the shore as she walks it in her light leaping stride, pocketing pieces of seaworn granite and quartz to be arranged on her tables outside with no purpose but every intention. Years ago, in 1954, when she was typing out Beckett's *Molloy* for the *Paris Review*, she became fascinated with the 'sucking stones' section.

'I spent some time at the seaside, without incident,' says Molloy. 'Personally I feel no worse there than anywhere else ... And to feel that there was one direction at least in which I could go no further, without first getting wet, then drowned, was a blessing.'

He then performs a strange, obsessive rite.

'I took advantage of being at the seaside to lay in a store of sucking-stones. They were pebbles but I call them stones. Yes, on this occasion I laid in a considerable store. I distributed them equally between my four pockets, and sucked them turn and turn about.'

'For ten pages, in one paragraph,' says Pat, 'he moves these stones in and out of his pockets and his mouth, working on a complicated logistic with the order of sucking each stone and where to put it after it is sucked so it won't get sucked again before all sixteen stones have, in turn, been sucked and put in the proper pocket. It took me a long time because I constantly got lost. I read and read this piece. Those stones stay with me . . .'

Stones and sea and sand. It's the nothingness of what she does that drives Pat on. Her energy has become concentrated, as if everything was working to some Zen-like point of absolute and discard; the apparent nothingness of her paintings, the seeming emptiness of the beach; as if she has conjured it all up herself, and is content with what she has done. She needs to do no more. Pat rarely leaves Provincetown now; she is bound to this place. 'I feel very cut off,' she said in 1987, more than twenty years after Nanno died. 'Come April, after a winter alone, I almost feel I don't exist.'

Living behind her trees, looking out to sea, she might be a forgotten figure in this forgetting town, abandoned all over again. But when we get in a taxi, the young driver tells me, 'Mrs de Groot rides for free.'

It lies there in the shadow of the wharf, as if it had sought shelter beneath the wooden struts. It has been dead for only twenty-four hours, but its distinctive markings – delicate grey and yellow swirls, merging as a graphic equaliser of its motion through the waves, as if they'd left their traces on its body – are already fading in the wind.

A common dolphin, exquisitely ill-named. Dennis writes the binomial down on his form, losing patience as his pen runs out: *Delphinus delphis*, a much more princely title, redolent of Cretan friezes and Greek vases. Two thousand years ago in his *History of Animals*, Aristotle attested to 'the mildness and gentleness of dolphins and the passion of their love for boys', and added, 'It is not known for what reason they run themselves aground on dry land; at all events it is said that they do so at times, and for no obvious reason.'

This is no wild strand on the Cape's ocean shore. It's the town beach on the bay, overlooked by the rear porches of shops and restaurants; this stranded cetacean might well have been a late-night throwaway, along with the lobster and clam shells. Yet these tame waters can be dangerous places, too. One morning, out on my deck, I'd seen fins in the distance, between the breakwater and the pier. Through my binoculars I watched a small pod of common dolphins moving restlessly up and down. I cycled down to see them from close quarters. Too close, I realised; they were in danger of grounding. I stood barely ankle-deep, and they were only twenty feet from me, where the blue became sandy brown. It seemed impossible that they could even be swimming there. The potential for disaster turned it into a quiet crisis, a clip from a natural history documentary with the voiceover removed, a scene ignored by the townsfolk going about their business.

For a dolphin to beach itself is a drastic act. Recent studies suggest that the animals 'will strand themselves when they are very weak because they don't want to drown', says Andrew Brownlow, a Scottish scientist. There seems to be 'something very deep in the terrestrial mammalian core that fires up when they are in extremis'. It is both suicidal and a desperate last attempt at survival. At least, that is how we see it. We sanctify these creatures as salves for our own depredations, and seem always to have done so. Around AD 180, the Greco-Roman poet Oppian declared that hunting 'the kingly dolphin' was immoral, on the grounds that they were once humans who had exchanged the land for the sea. 'But even now the righteous spirit of men in them preserves human thought and human deeds.'

Dennis called me with the news. Minutes later, we were driving down to the harbour. The day before, on the whalewatch boat, we'd watched the pod of dolphins moving through the clear waters in search of food. Among them was this individual. Such small groups of dolphins have close matrilineal relationships and are intensely loyal. Did it die in the night, on the dark and lonely beach, calling for its family as they called back? This beautiful, naked animal, now lying at my bended knees, was as smooth and patterned as a piece of porcelain. There was nothing morbid about it; it still seemed full of life.

I run my hands over its body. The fins are finely shaped, rubbery and tactile, caressed and caressing when alive; the taut flanks taper to the muscular tail. The eyes are disconcertingly open, unseeing, untouched by the gulls, which often fall to feed on stranded cetaceans even before they've expired. Clearly displayed on its underbelly is the animal's genital slit, flanked by two smaller mammary slits, betraying, in this indecent exposure, its sex. I insert my finger, ostensibly to investigate if she, as she had now become, had bred, but in reality out of prurient curiosity.

I say a Hail Mary for my sins.

After we have recorded her dimensions as if measuring her for a new outfit, I stretch out beside her for comparison; not for scientific reasons, but my own: head to tail, toe to beak, sensing how similar we are. I imagine her as a human in a dolphin wetsuit. I think of her bones, lighter than mine since they did not have to bear the full weight of gravity; I might replace my burdensome skeleton with hers, transformed from the inside out. I think about how much of my life is spent vertical or horizontal, upright on land or level with the water – a sensation known as proprioception: the apprehension of one's body in space; the way we want to be comfortable in the world, yet are never really reconciled to the business of being physical.

I lie there like a lover, her body a mirror for my own. Her blow-hole would never again burst open in exultation, in the joy of being

a dolphin. She wouldn't wriggle free of the sand, working her vigorous tail to swim away. The patina of decay had spread along her flanks like the silvery bloom on a plum. Dennis's knife cuts into the dorsal fin as the instructions on his form dictate, slicing off its tip in a liquorice-allsort sandwich of black skin and white fat. I feel an odd compulsion to bite down on the excised morsel. The teeth come next, each ivory needle arranged regularly along the narrow jaws. Research suggests that they may act as a sonic tool, helping to transmit sound back to a dolphin's inner ears.

Compared to this complex animal, I am sensorially inept, a dumb being barely able to feel anything. She could hear-see in the depths, heat-seeking sand eels and surfing with humpbacks; she could bond with her pod, using her signature whistle and those of her friends to call them. She could echo-locate her peers, sensing their emotional states, knowing how they felt, almost telepathically. She had a culture and expressed her self in a state of collective individualism and, as we now know, exhibited an emotional maturity possibly in excess of our own. But her life of apparent ease has been brought to an end on this urban shore. Passersby ask, 'What kind of fish is that?' Waiters sit on restaurant steps smoking cigarettes before the start of their next shift. In another age, their counterparts might have served it to their customers. In the nineteen-sixties, the town's Sea View diner had humpback on the menu.

Dennis saws at the jaw, hacking out the four teeth required for analysis by the organisation for whom he is acting. The serrated blade grates against bone, the worst hour in the dentist's chair you could imagine. The gums part and, two by two, the teeth are extracted. Blood trickles into the sand. The outrage is complete. Our samples bagged and the animal's flanks duly marked with the organisation's acronym, we drive off, leaving her alone on the beach, ready to roll in the next tide, as though its comforting waves might wash her back to life.

Dead or alive, we all strike the same pose; the same way my mother sat in a sepia photograph of her as a young woman in the garden of her suburban family home, resting her weight on one hand on the chair as she half turns to the camera like the movie stars she'd seen; the same way she'd sit in the last photograph I took sixty years later of her in our garden barely a mile away, adopting the same position; the same pose that, I realise, I too take up as I sit and turn to a camera which is not there.

Out in the bay, the moored boats act as weather vanes, swivelling and turning with the direction of the wind. I look out from my deck to the horizon. It's my barometer. If it's straight, there'll be whalewatching today; if it's wavy and irregular, perhaps not. Today it is level. So we go to sea.

There's nothing so exciting as that rising feeling as the boat readies to leave the harbour, potent with the prospect of the day ahead. Even as it stands tethered to the wharf, *Dolphin VIII* is a vessel invested with its own momentum, as though it would leave whether or not anybody was on board; a great grinding mass of steel plates and engines whirring deep down below, a powerful industrial connection with the resisting churning water. As I board with its crew – the fisherman turned captain, the taciturn first mate, the poet naturalist, the East European galley staff with professional futures back home – I feel a perennial outsider, for all that I've been sailing on these same boats, watching the same whales, for fifteen years. No one is ever sure of their place here, no one quite secure: the crew only work if the weather is good and the punters are paying their wages. Weather, work, people, whales: it is all an uneasy alliance, a nervous contract drawn up on an inconstant sea, agreed by a common pursuit. At least, for those few hours.

After a long bitter winter, the Cape has come to life. As I peer down into the green water, the reason is clear: fields of silvery sand eels, roused in their millions from the sea floor by the sun and now pooling in wriggling tangles, turning this way and that as one mass, just below the surface. These slender fish supply an entire food chain; their arrival could equally herald the crowds that will soon teem through the town's streets.

Only half an hour out from the land, a frenzy is in progress. Northern gannets are plunging into the bait like white-and-yellow torpedoes. A raft of loons, with stiletto-sharp bills and freckled oil-green wings, are working the same source. Harbour porpoises roll through the waves; grey seals bob like bottles.

Suddenly, something much larger appears in the one-hundred-and-fifty-foot-deep water that runs right up to Race Point: the falcate dorsal of a fin whale. For all its size, its black back too big to belong to a mere animal, it too is feeding on fish barely bigger than my finger. A pair of minkes, more modest rorquals, bearing the same strangely pleated bellies, join in. Then, as the boat pushes out over Stellwagen Bank's great drowned plateau under the wide Atlantic sky, the ocean begins to erupt anew with the blows of dozens of humpbacks, back from their winter stay in the Caribbean.

Then we are upon them, along with a thousand white-sided dolphin, weaving in and out as the great whales trap the sand eels in their bubble nets, rising through the corralled fish with mouths open wide, throats like rubbery concertinas, pleats clattering with barnacles like castanets. Gulls perch on the whales' snouts to pick out titbits. And just when it seems the scene can sustain no more predators, a dozen more fin whales arrive, lunging on their sides, displaying the bristly baleen in their jaws.

In this moment of witness, nothing else matters. Passengers delete images to make room for new ones on their cameras. My friend Jessica sees a couple frantically pressing the trash button as one says, 'Dump the wedding ones.'

Up on the bright white fly bridge, we watch the performance. A pair of adult fin whales aim straight for us. Each of them sixty feet long, at least.

Hands tight to the wheel, our captain, Todd Motta, shouts, 'Whoa!' as the nearest whale sheers off our bow, surfing on its side to display its great white belly like some enormous salmon.

'I thought it was going to hit us,' says Todd.

As experienced as he is, he's momentarily shaken. The second largest animal on earth, normally betraying barely a tenth of its mass as it moves through the sea, has flashed its entire physical self at us, using our boat as a fish stop. We are an instrument as much as an engine of observation.

All around us, the humpbacks continue to feed. One of the whales called Springboard rolls over to swim for a while on her back, displaying her genital mound, a region so gathered about with barnacles that it must make life uncomfortable for her suitors.

'I've never seen that before,' says Dennis.

Or maybe he has; it's so difficult to tell. Are these the same whales we just saw? The boat rocks and I stagger as I hold on to the clipboard and the rubber-encased GPS, regaining my footing to read off the coordinates for the pink photocopied sheets.

70 degrees north 18 degrees west. Mn: 1/2.

A calf holds its tail out of the waves, its body perpendicular in the water column. It trembles with its own life, the way a young boy's body trembles in adolescence, quivering with hormones. Then it starts to smash up and down on the water.

'Are these new animals?' Dennis asks.

I've no idea. The boat has turned round on itself, leaving a green swirling trail in its wake. The animals rise again, mouths as open as birds' beaks. The passengers look over the railings, ecstatically, loudly excited or overcome with lassitude and boredom, in the way of all ordinary miracles. None of this is of any consequence, because it happens day after day. Only in the actual moment am I transported. Only then does it leave me, this sense that I am not really here at all. We shiver with life, and its alternative. Waiting to come out the other side.

A few days later, we sail out of the harbour on another sunny morning. In the wheelhouse, I lean over the broad counter covered in what looks like wood-effect Formica from a seventies kitchen, peering at the chrome-ringed dials, updated with computer displays of the underwater terrain and a green radar screen silently scanning a black sea. We have left the land and its safety. An adhesive label announces the instructions for Marine Distress Communications to be relayed on the Submersible Plus VHF radio. Stuffed behind the sticky cup-holders is the Weekly Payroll Sheet.

Everyone on the bridge is in a good mood, looking forward to the day. But as the depth gauge draws 206 feet, the outlook changes as abruptly as the ocean floor falls away beneath us. The land to our starboard – such as it is – has been submerged under a sea fret. It's as if the view had reached the edge of an old projected film, fading into fuzzy nothingness.

The boat sails straight into the mist and everything around us disappears. The land and sky vanish into one vast cloud; all we are left with are the few yards of water immediately around the boat. We're entirely isolated, wrapped up in damp cotton wool. One minute, holiday sun; the next, murky obscurity.

'How do you look for whales in conditions like this?' I ask Lumby – Mark Dalomba, our captain for the day.

His camouflage cap is pulled down over his eyes; he doesn't turn round as he talks to me.

'Cut off the engines and listen,' he says. 'For the sound of their blows.'

But today Lumby has assistance. Chad Avellar, another young fisherman of Azorean descent who could sail these waters in the dark, is ahead of us, and radios back what he is seeing. Lumby charts a course ahead; or rather, he follows his own instincts. He plays the sea like a pinball machine. Perched on his captain's seat, eyes always ahead, he stabs at the radar screen.

'See those blips?' he says, pointing at the luminous green blobs shaping and reshaping, coming together in one mottled mass, discrete from the sea clutter that the fish-finder produces when reflected by the waves. 'Those are the whales.'

Conditions deteriorate. The boat rolls with its weight and ours, lurching from side to side.

'Crappy weather on the way,' says Lumby.

We seem to be moving ever slower, dragged back by the banks of fog. My heart sinks. It's my last trip of the season. Even if we come upon whales, will we actually see them? Everything is grey.

There's no horizon, no context. We might as well have drifted into the Arctic, or the Bermuda Triangle, for that matter.

The silence explodes with blows. Of course it does. We are *surrounded* by whales, as if they'd been there all along, only now choosing to break cover. The water bursts with their exhalations. We can't tell sea from sky, but these animals are producing their own weather, their spouts merging with the mist.

They are feeding, voraciously. Bellowing, blowing, rising up through their own bubble-clouds, eight whales at a time piercing the surface, cooperating in an orgy of consumption. It is a visceral, indisputable, audible frenzy. Whales are not tentative. They do not fuss and bother. They do not falter. They act, uproariously, greedily, and utterly in-their-moment.

Lumby climbs up to the fly bridge. As he does so a dozen whales loom up right off the bow, their cavernous mouths open like gigantic frogs, fringed with baleen and roofed with pink strips like engorged tongues. It's a fearsome sight. We follow Lumby aloft, clambering up after our captain as if trying to get away from the beasts.

From our eyrie, we look down through the mist. Everywhere there are whales, lunging and fluking and kick-feeding, taking advantage of the fog to cover their gluttony. Fifteen humpbacks, maybe more.

Then, as if roused by their mothers' feeding frenzy, the calves begin to leap. One after another, spindle-shaped bodies shoot out of the sea like popguns going off. We don't know where to look. Lumby holds the boat in position; he seems to be conducting the whole scene, even though he has lost control, like the rest of us.

'Jesus Christ,' I exclaim, then apologise, hoping the passengers haven't heard me.

'No,' says Liz, the poet naturalist. 'That's quite appropriate.'

The calves have begun to breach simultaneously: two, three, four, five, all together.

'They're more like dolphins than whales,' I shout.

No marine park could rival this show. They might as well be Eocene cetaceans leaping out of an ancient ocean, celebrating their leaving of the worrisome land. Two centuries ago, as a young man on his maiden voyage, Melville saw his first whales not far from this shore; his ship, too, was drifting in the mist.

'The most strange and unheard-of noises came out of the fog at times: a vast sound of sighing and sobbing. What could it be? This would be followed by a spout, and a gush, and a cascading commotion, as if some fountain had suddenly jetted out of the ocean . . . But presently some one cried out – "*There she blows! whales! whales close alongside!*"' To the young sailor, they sounded like a herd of ocean-elephants.

As the sea bursts with the blows and foraging of the adults, it is blown open by their breaching calves, creating abbreviated geyser-spouts of their own. Up on the bridge, we've run out of superlatives. John, our hardbitten first mate, is speechless. Later, in the afterglow of what we've witnessed, in a kind of apologetic embarrassment of emotion, he volunteers that, out of seven thousand trips, this is one to remember – 'And it takes a lot to impress me.' Liz and I assure our passengers – should they assume that this sort of thing happens every day – that it is one of the most extraordinary sights we have seen, out here on the Bank.

Then I look at Lumby. Under the peak of his cap, tugging at the cigarette jammed in his fist, he too is smiling to himself, as if he had summoned it all up. As if the scene, all the more amazing for the inauspiciousness of its prelude, were a vindication of his magical skills, far beyond those of naturalists or scientists or writers. Like his fellow captains, Lumby has never taken a photograph of a whale.

He doesn't need to. They're all there, in his head.

THESTARLIKESORROWS OFIMMORTALEYES

I return to the Cape on the eve of the new year. The summer is long gone. The sun looks as strong as ever, but it is made milky by the cold, its span over the horizon shortened. The days open late, become public, flicker, then close early, reclaiming their privacy.

As Dennis and Dory and I walk the beach at Herring Cove, the Arctic wind hits us full on. It bites at my face, tearing off the sun's facile heat. I pull my scarf over my nose and stumble through the sand. Dennis kneels to the ground; we observe the rituals of the dead. A herring gull lies eviscerated, its guts pecked out by a glaucous gull which we saw at a distance, crouching over its cousin, ready and welcome protein. Dennis records the carcase on an index card. The blood, on pure white feathers, is strangely orange. The hole in its belly is big enough for me to wear the dead bird as a hat, should I so wish.

I throw Dory's ball. She is naked, save for her collar. I worry that she might be shivering too. Her brow furrows and she cocks her head to one side as she asks me to throw the ball again. When we are with dogs, physicality is uncomplicated. They walk beside us as our outliers. Part of the human party, they are also our bridge with the natural world. They are our other. They are not cleverer than us, so we love them.

Like all animals, Dory has extraordinary eyes. Hers are fringed with pale lashes. No human could look so exquisite, or so feral, so unadorned. I can see why people once worshipped dogs. As we drive to the beach Dory perches on the armrest between Dennis and me, peering intently ahead, seeing and hearing things we do not see or hear. We only know because her ears rise or her eyes twitch. She knows where we are going. Perpetually expectant, as if every experience were a surprise, her body quivers with the excitement of just being alive. It is *Funktionslust*; an animal's pleasure in doing what it does well, in being itself.

Dory is an import, like everyone else here, rescued from the backstreets of Miami. Now she scents foxes and chases balls, sometimes letting them roll into the surf, then staring at them as if daring me to go in after them. Her breeding, such as it is, may be Caribbean – a wild dog, the sort you see roaming Haitian beaches in packs and howling in the heat of the night – but her compact body seems suited to this winter landscape. Her neat flat coat is the colour of the dunes and the parched grass, although now she is growing fine silver hairs through her desert pelt. She never stops being, never stops running for her ball; I think her heart would burst before she let up the chase. Her life runs ahead of ours, speeding up as she races alongside us in another time zone. I'd like to talk to her in her voice, but like Wittgenstein and his lion, I wouldn't understand what I might hear. Debbie, Dennis's wife, says that sometimes Dory comes back from the woods shaking as if in fear, as if she'd seen something out there.

'I am secretly afraid of animals,' Edith Wharton, an erstwhile New Englander for all that she spent almost all her life in Paris, wrote in 1924, '– of all animals except dogs, and even of some dogs. I think it is because of the *Usness* in their eyes, with the underlying *not-usness* which belies it, and is so tragic a reminder of the lost age when we human beings branched off and left them; left them to eternal inarticulateness and slavery. *Why?* their eyes seem to ask us.'

The wonder is that all animals are not afraid of us. J. A. Baker, who spent the nineteen-sixties observing the wildlife of Essex, wrote of finding a heron on the winter marshes, trapped by its wings frozen to the ground. Baker dispatched the bird, humanely, watching the light leave its frightened gaze and 'the agonised sunlight of its eyes slowly heal with cloud'.

'No pain, no death, is more terrible to a wild creature than its fear of man,' Baker concludes, in a deeply affecting passage, cited by Robert Macfarlane: 'A poisoned crow, gaping and helplessly floundering in the grass, bright yellow foam bubbling from its throat, will dash itself up again and again on the descending wall of air, if you try and catch it. A rabbit, inflated and foul with myxomatosis ... will feel the vibration of your footstep and will look for you with bulging, sightless eyes. We are the killers. We stink of death.' Nature writing becomes war reporting. I remember the countryside of my childhood infected with that disease. In his 'Myxomatosis', written in 1955, Philip Larkin sees a rabbit 'caught in the centre of a soundless field', and uses his stick in an act of mercy. 'You may have thought things would come right again | If you could only keep quite still and wait.' My sister remembers our father having to do the same thing: the same terrifying eyes, the same dispatch.

We only play our roles; animals' fates are our own. The fifteenth-century orator Pico della Mirandola, in his essay 'On the Dignity of Man', declared that to be human is to be caught between God and animal: 'We have set thee at the world's centre that thou mayest more easily observe what is in the world.' Five hundred years later, the Caribbean writer Monique Roffey saw that 'Animals fill the gap between man and God.' That gap has widened. As John Berger observed, animals furnished our first myths; we saw them in the stars and in ourselves. 'Animals came from over the horizon. They belonged there and here. Likewise they were mortal and immortal.' But in the past two hundred years, they have gradually disappeared from our world, both physically and metaphysically: 'Today we live without them. And in this new solitude, anthropomorphism makes us doubly uneasy.'

We expect animals to be human, like us, forgetting that we are animals, like them. They 'are not brethren, they are not underlings', the naturalist Henry Beston wrote from his Cape Cod shack in the nineteen-twenties; to him, animals were 'gifted with extensions of the senses we have lost or never attained, living by voices we shall never hear ... other nations, caught with ourselves in the net of life and time, fellow prisoners of the splendour and travail of the earth'. That fear we see in their eyes is fear in alien eyes, eyes created for other realms.

Dennis, Dory and I walk on, around the cove. The tide peels back time, revealing frozen expanses of sand and waves of wrack. Skeins of briar and line have twined together like an elongated net constructed to catch primeval fish. I half expect to see a Neolithic family foraging on the beach. The landscape is moonlike, bone-scattered. Sere, stripped back by the winter, pallid and raw. Yet despite the intense cold – so barbarous it becomes a kind of warmth – the shore is full of life.

Everything is residual and tentative in the intertidal zone; a place belonging to no one, 'a sort of chaos', as Thoreau saw it, 'which only anomalous creatures can inhabit'. Ribbed mussels, elegantly slipper-shaped in metallic blue and mauve, lie next to tiny flat stones, beige and green and purple and ringed with white. Through this tesserated pavement, samphire pushes up its stiff fingers; it's called pickle grass here, a name that sums up its salty gherkin crunch. Stattice stands upright; even its everlasting purple flowers have been leached to a lifeless brown. Wind-burnt stalks of wild rose have long since lost their scent, but can still tear bare flesh. Pale-green lichens, barely alive at all, grow infinitesimally, stone flowers in this tundra-by-the-sea.

Dennis shows me his favourite tree: a stunted cedar like a large bonsai, spreading its skirt over a sandy hillock, as though claiming the site of an ancient tumulus. Impaled in a bayberry bush is the empty shell of a crab, probably dropped by a passing gull, still snapping its upraised claws at the sea across the dunes.

The estuary ahead widens with the falling tide. In the distance is Race Point Light. In between is Hatches Harbor, site of another lost settlement, like Long Point. Dennis thinks this was the place known as Helltown – an outpost for the outcasts of an already remote place, the human reverse of this heaven. Perhaps it resembled Billingsgate Island down at Wellfleet, which was reserved for young men, with its own whale-lookout, tavern and brothel.

Today there's not a soul to be seen on this beach. But one winter morning I arrived here to see what looked like black sails a mile down the shore. As they dipped and swayed, I thought they belonged to particularly intrepid windsurfers. Only when I raised my binoculars did I see that the dark triangles were rising and falling, flexingly powered by something far bigger and stronger than a wetsuited human. I realised, with a sharp intake of cold breath, that they were the flukes of right whales, rolling in the waves.

Trying to remember the intricate geography of this outermost edge of the Cape, I cycled round to the fire road and as far as I could to the distant beach, abandoning my bike in the dunes. I would have run if the sand had allowed me. Cresting a low hill and stumbling through the marran grass, I suddenly regained the shore.

Below me lay a great crescent-shaped arena, occupied by hundreds of herring gulls. As I approached, they rose as one like a theatre curtain to reveal, barely twenty yards beyond the surf, half a dozen right whales engaged in what scientists call a surface active group, and what you and I might call foreplay.

I crouched there, doing my best not to disturb them. For an hour or more I watched their sleek blubbery bodies tumble and turn over one another in an intimate display, all the stranger and more physical for their nearness to the shore, as if they might be beached in the throes of their passion. But nothing could have curtailed those caresses. A harbour seal sat at the waterline watching too, hesitating to share the waves with these loved-up leviathans. It was a spectacle made more extreme by the cold, the sun, the wind and the silence as these gigantic animals, whose glossiness seemed to absorb all the light and the energy of the day, danced around one another in an amorous ballet whose choreography was determined only by their own sensuality.

There are no whales today, amorous or otherwise. Perhaps it

is too cold even for their courtship. Dennis and I take shelter in the lee of a dune. For a few moments we're out of the wind and can draw warm breath again. With the sun on our backs, our muscles relax. Hunched shoulders and curled hands loosen a little. As I look around, I realise that we are surrounded by bones – femurs and sternums, ribs and skulls – all tangled up in the salt hay.

We're standing in a graveyard, an animal ossuary.

Poking about in the wrack we find a fox splayed in the tousled seaweed as if caught in the act of running, or of agony. Its flesh has been stripped away like an anatomical drawing. Clenched jaws display fine canines; ribs are picked clean. But its brush, the length of its body again, streams behind it, resplendent, rotting.

Nearby is a gannet. Or rather, its wings, six feet wide, great white-and-black contraptions discarded by some modern Icarus who'd fallen face-first into the grainy wet sand, leaving a pair of feet sticking out of the ground. A single gannet would fill my box bedroom back home; a bird on a giant scale. I hold the feathers up behind my back, as if the fledgling buds had burst through my skin, sprouting from my shoulderblades and unfolding to lift me into the air. I remember reading in my children's encyclopaedia that my dreams of growing functional wings were impossible, because I'd have to grow a breastbone longer than my body. The accompanying illustration showed a man with his sternum hanging down between his legs, like some grotesque man-bird chimera drawn by Leonardo.

We turn back into the wind. The strand sweeps open and wide, connecting the inner bay with the outer sea. A beach that in summer is filled with sunbathers and anglers remains resolutely empty. I take off my clothes – no easy matter with frozen fingers and gloves, hat, scarf, jacket, fleece, two jumpers, boots and socks and jeans and long johns to contend with – and run into the navy-blue sea. It rolls on, and on. It looks like it did six months ago and five thousand years ago. It even feels the same. I treat it accordingly, borne up, singing, as though nothing has changed. As though everything will always be like this, and always was.

It is New Year's Day.

Dennis and Dory walk on ahead. Glowing pink and shivering like a dog, my extremities as navy blue as the sea, I struggle back into my clothes, unable to do my jacket up with my numb fingers, and run after them. Dory looks back, apparently relieved. Did she think I'd been lost for good? In the car park Dennis has to rub my hands in his, making jokes about hoping that none of his friends will be driving by. My teeth chatter and my muscles shiver, shaking me back to life. Skin and bone burn like a hard cold flame. And they continue to burn and shake for an hour afterwards, till my body is convinced that the threat is over. Every swim is a little death. But it is also a reminder that you are alive.

Out at sea, hundreds of eiders and mergansers bob in the waves. They must be among the most hardy of all animals, these sea ducks, forever riding on the freezing water, resilient and resigned. At the north end of Herring Cove – in the lee of the rip of the Race where the sea turns dark as it becomes the ocean – is a sandbar which traps a temporary lagoon at high tide. In the heat of summer it's a wonderfully warm place to swim, as languorous as a Mediterranean pool, although once I was horrified to see half a humpback beneath me, its great white knobbly flipper all but waving to me from the sandy bottom, as if its part-carcase were preserved by the salt water. Today the tide is running fast, and would quickly carry me out to sea.

This entire rounded tip of the Cape is a curling catchment, a beneficiary of long shore drift, perpetually shifting to reveal shipwrecks sticking out of the dunes. After winter storms have destroyed most of the car park – leaving its tarmac hanging in slabs like cooled lava over the sand – a chunk of ship, stirred out of retirement, emerges up the beach. Was it washed up or merely uncovered by the storms, lying there all along as I walked there, its knees and ribs beneath me, rubbed and eroded by the decades in which they have rolled around on the sea bed, waiting to be revealed like some vast wooden whale? It might be the remains of a twentieth-century vessel or a Viking longboat. The splintered timbers and curled ribs of oak lie cloaked in emerald-green weed, bolted pieces of something whose shape can only be guessed at.

'The annals of this voracious beach! who could write them, unless it were a shipwrecked sailor?' Thoreau wrote as he wandered from one end of the Cape to the other from 1849 to 1857, continually drawn back to this inbetween place. 'How many who have seen it have seen it only in the midst of danger and distress, the last strip of earth which their mortal eyes beheld! Think of the amount of suffering which a single strand has witnessed! The ancients would have represented it as a sea-monster with open jaws, more terrible than Scylla and Charybdis.'

Walking towards Provincetown, Thoreau saw an arrangement of bleached bones on the beach ahead, a mile before he reached them; only then did he realise they were human, with scraps of dried flesh still on them. It was a sign Shelley had already foreseen, 'On the beach of a northern sea', as if in a premonition of his own demise, 'a solitary heap, | One white skull and seven dry bones, | On the margin of the stones'.

On another walk Thoreau was told of two bodies found on the strand: a man, and a corpulent woman. 'The man had thick boots on, though his head was off, but "it was along-side". It took the finder some weeks to get over the sight. Perhaps they were man and wife, and whom God had joined the ocean-currents had not put asunder.' Like the victims of *Titanic*, some bodies were 'boxed up and sunk' at sea; others were buried in the sand. 'There are more consequences to a shipwreck than the underwriters notice,' said Thoreau. 'The Gulf Stream may return some to their native shores, or drop them in some out-of-the-way cave of ocean, where time and the elements will write new riddles with their bones.' I see that same sea in his eyes, eyes that seem to see the sea forever; what it had found, and what it had lost.

Nearly four thousand ships have been wrecked along the Cape's outer shore, from *Sparrowhawk*, which ran aground down at Orleans in 1626 and whose survivors were given refuge by the Pilgrims at Plymouth, to the British ship *Somerset*, which came to grief off Race Point in 1778 during the Revolutionary War, having fought in the Battle of Bunker Hill, foundering on the sandy bar off the Race. Twenty-one of its sailors and marines drowned, but more than four hundred were taken prisoner and sent to Boston. The Cape Codders escorting them gave up halfway, possibly worn down by their charges asking, 'Are we there yet?' *Somerset* has appeared every century since, in 1886, 1973 and 2010; a spirit ship, a beached *Flying Dutchman*. One writer in the nineteen-forties claimed that scores of people had seen 'ghosts in the vicinity, ghosts of the British sailors'. The ship remains a sovereign vessel; perhaps I ought to reclaim it for my queen.

Meanwhile, many other wrecks lie out there like time machines. Thoreau saw the bottom of the sea as 'strewn with anchors, some deeper and some shallower, and alternately covered and uncovered by the sand, perchance with a small length of iron cable still attached, – of which where is the other end?'

'So many unconcluded tales to be continued another time,' he wrote. 'So, if we had diving-bells adapted to the spiritual deeps, we should see anchors with their cables attached, as thick as eels in vinegar, all wriggling vainly toward their holding-ground. But that is not treasure for us which another man has lost; rather it is for us to seek what no other man has found or can find.'

Wreckage and the wrecked: they merge into one, a mangling of man and land, of vessel and sea. I think of Crusoe cast up on the shore waiting for Friday's footsteps, as the waves washed over a plaintive nineteen-sixties soundtrack; of Ishmael, another orphan, clinging to a coffin carved for Queequeg which provided his lifebuoy; of beached whales and beached humans. And I hear my father singing, 'My bonny lies over the ocean, my bonny lies over the sea, my bonny lies over the ocean, O, bring back my bonny to me.' I used to hear 'body' for 'bonny'.

When Thoreau was visiting the Cape, an average of two ships every month would be lost in winter storms, especially on the deceptive bars off the Race at Peaked Hill, where shoulders of sand shadow the ocean's edge. The roaring breakers catch white on the shifting shelf, luminous at night with the memory of the lives they've taken. And all this happened within sight of land.

'Ship ashore! All hands perishing!'

These tempests were not conjured up by a magician, nor were there any sprites on hand to guide the survivors to safety. Commonly, sailors did not learn to swim – partly through superstition – 'What the sea wants, the sea will have' – and partly through practicality, knowing that adrift on the open ocean, their flailing would only prolong their fate. Any attempts to save the shipwrecked were often defeated by the elements. Would-be rescuers could only look on and wait until the storms subsided, by which time it was too late. All that was left to do was to salvage the wreck. In the eccentric museum at the Highland Light, housed in a 1906 hotel standing in the shadow of the lighthouse on the windblown headland, one of the most haunted places I have ever visited, a row of assorted chairs from many different disasters stands as a testimony to lost souls and salvaged domesticity: a sad line of mismatched seating, ranged along a wall at a students' party. Upstairs, rooms with stable doors lined along a long, narrow and dimly-lit corridor; they still seemed filled with fitful guests, and something in the darkness down the end told me to get out.

Those who did make it ashore could die of exposure in this no-man's-land, with no hope of reaching dwellings set deep inland, far from the raging sea. In 1797 the Massachusetts Humane Society set up 'Humane-houses', a series of huts equipped with straw and matches to provide survivors with warmth and shelter. Their echoes remain in the shacks still scattered through the dunes: rough constructions put together from grey beach-wood and timbers as though assembled by those lost sailors. Even in the town, salvaged ships' knees propped up houses against the storms that brought the flotsam here, while Thoreau recorded fences woven with whale ribs.

Other dangers lurk in these countervailing waters, seen and unseen. Locals told Thoreau there was 'no bathing on the Atlantic side, on account of the undertow and the rumour of sharks', and he was warned by the lighthouse keepers at Truro and Eastham not to swim in the surf. They would not do so for any sum, 'for they sometimes saw the sharks tossed up and quiver for a moment on the sand'. Thoreau doubted this, although he did see a six-foot fish prowling within thirty feet of the shore. 'It was of a pale brown color, singularly film-like and indistinct in the water, as if all nature abetted this child of ocean.' He watched it come into a cove 'or bathing-tub', in which he had been swimming, where the water was just four or five feet deep, 'and after exploring it go slowly out again'. Undeterred, Thoreau continued to swim there, 'only observing first from the bank if the cove was preoccupied'.

To the philosopher, this back shore seemed 'fuller of life, more aërated perhaps than that of the Bay, like soda water', its wildness lent an extra charge by that sense of life and death. Down at Ballston Beach, where Mary and I often swim out of season while seals and whales feed just off the sandbar, the powerful undertow seeks to pull us out. Not long ago a man swimming here with his son was bitten by a great white shark. Public notices instruct swimmers to avoid seals, the sharks' true targets. Recently a fisherman showed me a photograph on his phone, taken at Race Point. A great white breaks the surfline, barely in the water, with its teeth around a fat grey seal. I place my quivering body in that tender bite, the 'white gliding ghostliness of repose' which Ishmael discerned, 'the white stillness of death in this shark'. I still swim there, despite Todd Motta's warning, 'You don't wanna go like that.' The water is as hard and cold as ever. But one day, I think, I will not come out of it.

In his book *The Perfect Storm*, the story of a great gale which hit New England in 1991, Sebastian Junger details the way a human drowns. 'The instinct not to breathe underwater is so strong that it overcomes the agony of running out of air.' The brain, desperate to maintain itself to the last breath, will not give the order to inhale until it is nearly losing consciousness. This is the break point. In adults, it comes after about eighty seconds. It is a drastic decision, a final, fatal choice – like an ailing dolphin deciding to strand rather than drown because of something deep in its mammalian core; 'a sort of neurological optimism', as Junger puts it, 'as if the body were saying, *Holding our breath is killing us, and breathing in might not kill us, so we might as well breathe in.*'

Drawing in water rather than air, human lungs quickly flood. But lack of oxygen will have already, in those last seconds, created a sensation of darkness closing in, like a camera aperture stopping down. I imagine that receding light, being drawn deep, caught between the life I am leaving and the eternity I am entering. We know, from those who have come back from death, that 'the panic of a drowning person is mixed with an odd incredulity that this is actually happening'. Their last thoughts may be, 'So this is drowning,' says Junger, 'So this is how my life finally ends.'

And at that final moment, what? Who will take care of my dog? What will happen to my work? Did I turn the gas off? 'The drowning person may feel as if it's the last, greatest act of stupidity in his life.' One man who nearly drowned, a Scottish doctor sailing by steamship to Ceylon in 1892, reported the struggle of his body as it fought for the last gasps of oxygen, his bones contorting with the effort, only to give way to a strangely pleasant feeling as the pain disappeared and he began to lose consciousness. He remembered, in that instant, that his old teacher had told him that drowning was the least painful way to die, 'like falling about in a green field in early summer'.

It is that euphoria which offers an aesthetic end, leaving the body whole and inviolate, a beautiful corpse, as if the sea might preserve you for eternity. There is an inviting compulsion about falling into the sea, because it seems such an unmessy, arbitrary way to go. You're there one minute, in another world the next; a transition, rather than a destruction.

On his journey from New York to England in 1849 on the ship *Southampton*, Melville saw a man in the sea. 'For an instant, I thought I was dreaming; for no one else seemed to see what I did. Next moment, I shouted "Man overboard!"' He was amazed that none of the passengers or sailors seemed very anxious to save the man. He threw the tackle of the quarter boat into the water, but the victim could not, or would not, catch hold of it.

The whole incident played out in a strange, muted manner, as if no one really noticed or cared, not even the man himself.

'His conduct was unaccountable; he could have saved himself, had he been so minded. I was struck by the expression of his face in the water. It was merry. At last he drifted off under the ship's counter, & all hands cried, "He's gone!"'

Running to the taffrail, Melville watched the man floating off, 'saw a few bubbles, & never saw him again. No boat was lowered, no sail was shortened, hardly any noise was made. The man drowned like a bullock.'

Melville learned afterwards that the man had declared several times that he would jump overboard; just before his final act, he'd tried to take his child with him, in his arms. The captain said he'd witnessed at least five other such incidents. Even as efforts were made to save her husband, one woman had said it was no good, '& when he was drowned, she said "there were plenty more men to be had."'

Half a century later, in 1909, Jack London – who was a deep admirer of Melville – published *Martin Eden*, his semi-autobiographical account of a rough young sailor who becomes a writer. London, the son of an astrologist and a spiritualist, was born in San Francisco in 1876. He had led an itinerant life as seaman, tramp and gold prospector. He was a self-described 'blond-beast', a man of action, the first person to introduce surfing from Hawaii to California; he also became the highest-paid author in the world with books such as *The Call of the Wild*, *The Sea-Wolf* and *White Fang*. But the proudest achievement of his life, he said, was an hour spent steering a sealing ship through a typhoon. 'With my own hands I had done my trick at the wheel and guided a hundred tons of wood and iron through a few million tons of wind and waves.'

London wrote *Martin Eden* while sailing the South Pacific, trying to escape his own fame; the *New York Times* had reported FEAR JACK LONDON IS LOST IN PACIFIC when he didn't arrive as expected at the Marquesas, the remote islands where Melville himself had jumped ship in 1840. In London's book, Eden, the first man to himself, cynical about his new-found celebrity, contemplates suicide in the early hours of the morning. He thinks of Longfellow's lines – 'The sea is still and deep; | All things within its bosom sleep; | A single step and all is o'er, | A plunge, a bubble, and no more' – and decides to take that step. Midway to the Marquesas, he opens the porthole in his cabin and lowers himself out.

Hanging by his fingertips, Eden can feel his feet dangling in the waves below. The surf surges up to pull him in. He lets go.

Everything in his strong constitution fights against this act of self-destruction. As he hits the water, he begins to swim; his arms and legs move independently of his will, 'as though it were his intention to make for the nearest land a thousand miles or so away'. A tuna takes a bite out of his white body. He laughs out loud. He tries to breathe the water in, 'deeply, deliberately, after the manner of a man taking an anaesthetic'. But even as he pushes his body down vertically, sinking like 'a white statue into the sea', he is pushed back to the surface, 'into the clear sight of the stars'.

Finally, Eden fills his lungs with air and dives head-first, past luminous tuna, plunging as deep as he can. His body bursts with bubbles. He is aware of a flashing bright light, like a lighthouse in his brain. He feels he is tumbling down an interminable stairway. 'And somewhere at the bottom he fell into darkness. That much he knew. He had fallen into darkness. And at the instant he knew, he ceased to know.' Eden, this handsome sailor whose body is described as solid, hewn and tanned, is compressed by the weight and darkness of the sea, falling asleep on its bed, so still that a shake of the shoulder could not wake him. He has been sacrificed to his own ideals, his own masculinity. London said his novel was about a man who had to die, 'not because of his lack of faith in God, but because of his lack of faith in men'. His writing is vivid enough to recall his own youthful attempt to drown himself in San Francisco Bay, when 'some maundering fancy of going out with the tide suddenly obsessed me'. 'The water was delicious,' he wrote. 'It was a man's way to die.'

There have been moments in the water when I felt they might be my last. One dark November afternoon I swam off Brighton, under the shadow of its burnt-out West Pier, while a murmuration of starlings eddied about the rusting ribs above me. I hadn't realised, until I entered the water, how strong the undertow was; or, as I swam out, how it would take me up, take control, tipping me head over heels before dragging me back out.

I'd lost my grip on the world. The heavy pebbles of the beach rolled beneath me, and in the falling darkness, as the lights came on along the esplanade, I thought how banal it would be to die within sight of a dual carriageway and a row of fish-and-chip shops and burger bars. And I wonder, when I am dead, what thoughts will be left in my head, like the black box recorder of a downed plane.

Another time, on Dorset's West Bay, under its towering cliffs, the tow played a similar trick. I quickly realised what I had done, and tried to climb out. Again I was turned over for my impudence and thrown face-down on the shingle, my features squashed like a peat bog man. Mark told me this was the way surfers smashed their faces, and that evening in town, someone warned me that the beach was notorious, and that only a few months before a young man had drowned there.

And I thought about Virginia Woolf's body being taken out, as if her death were a culmination of all her words, moving inexorably towards the sea.

It's odd to return to the books I was required to read at college, their unbroken backs covered in clear plastic to protect them against some future event, preserving them for a time when I would actually understand them, although their pages are now vignetted in brown, as if the sun had penetrated their closed edges. They wait for me to open them, to bring them back to life, familiar and strange and dangerous, as though I were reading them for the first time.

To the Lighthouse is set in the Hebrides, but it draws on Woolf's childhood holidays in Cornwall, and memories of her Victorian mother. Mrs Ramsay hears and feels the waves as they 'remorselessly beat the message of life'; they make her think of 'the destruction of the island and its engulfment in the sea'. At night, as her guests sit around the candlelit table, she looks out of the uncurtained windows through the dark rippling glass – 'a reflection in which things waved and vanished, waterily', as if all the world was at sea – and she thinks of herself as a sailor who, if the ship had sunk, 'would have whirled round and round and found rest on the floor of the sea'. In the distance, the lighthouse stands tall and white on a rock.

The water possessed an ambivalent power for Woolf. One moonlit night, when she was a young woman, she and Rupert Brooke swam naked in the river Cam at Byron's Pool, named after the poet, who had swum there when he was at Cambridge. Brooke was proud of his improbable and Byronic ability to emerge from the water with an erection. Later, Woolf joined Brooke and his Neo-Pagans, as she called them, when they camped on Dartmoor and swam in the moorland river. Virginia, both prim and liberated, did not quite feel at ease with their attempts to commune with nature; her future biographer Hermione Lee would lament the fact that the nude photographs taken on that occasion did not survive.

Woolf – only an extra O away from being an animal herself, a virgin wolf – had a relationship with the natural world that was both paradoxical and predatory. Nature was unfeeling, going about its business. The beach was no consolation. In *To the Lighthouse*, after a scene in which 'the sea tosses itself and breaks itself, and should any sleeper fancying that he might find on the beach an answer to his doubts, a sharer of his solitude, throw off his bedclothes and go down by himself to walk on the sand ... to ask the night those questions as to what, and why, and wherefore, which tempt the sleeper from his bed to seek an answer' – we discover, almost in passing, that Mrs Ramsay has died. In the aftermath, the sea seems to take over the house, as death has overtaken the Ramsays. Of their eight children, Andrew is killed in the war and Pru dies in childbirth. Virginia's own mother, Julia, died aged forty-nine, and her brother Thoby died of typhoid fever when he was twenty-six years old. For Woolf, the water meant death as well as life.

What remains of the Ramsay family and their friends return ten years later. The house, once so full, has stood empty; the elements threaten to overtake it. We expect the deluge of war to have washed it away. But it is rescued by the housekeeper, to whom Mrs Ramsay appears as a 'faint and flickering' image, a kind of ghost, 'like a yellow beam or the circle at the end of a telescope, a lady in a grey cloak, stooping over her flowers'. The memory is electric, almost cinematic: Virginia's mother Julia was photographed by her aunt, Julia Margaret Cameron, more than fifty times, her profile turned this way or that, her smooth hair, glaucous eyes and strangely vacant face the same as her daughter's, wearing a black gown, white cuffs and collar, caught on the path at Freshwater, moving in her dark clothes; then not moving, stilled in the instant, then moving on, 'the Star like sorrows of Immortal Eyes'.

So too Virginia would pose for *Vogue* in her mother's dress in 1924, ravished by a Pre-Raphaelite sea, acting as her own sepia ghost, rehearsing her last scene, floating down the Ouse as Ophelia, 'her clothes spread wide, | And mermaid-like, awhile they bore her up'. After her father died, and Virginia and her orphaned siblings moved to Bloomsbury, she hung Cameron's fantastical portraits of famous men and fair women in the hallway as an ironic gesture. For all her modernism, Virginia was anchored in a Victorian past, shaped and damaged by its history, and her own.

Those remote summers by the sea would remain with her. In her book, the ferocious Atlantic becomes a character itself, like the moor in *Wuthering Heights* or the whale in *Moby-Dick* (of which she owned two copies, and which she read at least three times). 'In both books,' she wrote in an essay on Brontë and Melville in 1919, 'we get a vision of presence outside the human beings, of a meaning that they stand for, without ceasing to be themselves.' Woolf's white lighthouse is Melville's white whale; an impossible mission over unfathomable waters.

Cam, the riverishly-named youngest Ramsay daughter, dangles her hand in the waves as she and her brother reluctantly accompany their father on the long-postponed trip to the lighthouse. Out at sea they become becalmed, and in her dreamy, deceptive state, Cam's mind wanders through the green swirls into an 'underworld of waters where the pearls stuck in clusters to white sprays, where in the green light a change came over one's entire mind and one's body shone half transparent enveloped in a green cloak'. As blank and ever-changing as it is, as calm or crazed, the sea could embody ecstasy or despair; it was a mirror for Woolf's descent into madness, a process made profound by knowing what was about to happen. She might have been enchanted by Ariel. 'I felt unreason slowly tingling in my veins,' she would say, as if her body were being flooded by insanity or filled with strange noises: birds singing in Greek; an 'odd whirring of wings in the head'.

Cam seems besieged by the sea, by a numb terror and 'a purple stain upon the bland surface as if something had boiled and bled, invisibly, beneath'. Meanwhile 'winds and waves disported themselves like the amorphous bulks of leviathans whose brows are pierced by no light of reason'. Eventually the Ramsays reach the lighthouse, but even that epiphany is darkened by the fact that they pass over the place – if water could be said to have a place – where their fisherman had once seen three men drown, clinging to the mast of their boat. All the while their father, as gloomy and tyrannical as Ahab, dwells on William Cowper's doomy poem, 'The Castaway': 'We perish'd, each alone: | But I beneath a rougher sea, | And whelm'd in deeper gulfs than he.' When, as a young woman, Virginia had heard of the fate of Titanic, she imagined the ship far below, 'poised half way down, and become perfectly flat', and its wealthy passengers 'like a pancake', their eyes 'like copper coins'. Later, to another friend, she said, 'You'll tell me I'm a failure as a writer, as well as a failure as a woman. Then I shall take a dive into the Serpentine, which, I see, is 6 feet deep in malodorous mud.' To her even the bridge over the monstrously-named inland sea in a London park was a white arch representing a thousand deaths, a thousand sighs.

While writing To the Lighthouse, Woolf read of another disaster. On the first ever attempt to fly westbound across the Atlantic, the wealthy Princess Löwenstein-Wertheim had perished, along with her pilot and co-pilot. 'The Flying Princess, I forget her name, has been drowned in her purple leather breeches.' In her mind's eye Virginia saw the plane running out of petrol, falling upon 'the long slow Atlantic waves' as the pilots looked back at the 'broad cheeked desperate eyed vulgar princess' and 'made some desperate dry statement' before a wave broke over the wing and washed them all into the sea. It was an arch nineteen-twenties scene; Noël Coward out of The Tempest. 'And she said something theatrical I daresay; nobody was sincere; all acted a part; nobody shrieked.' The last man looked at the moon and the waves and, 'with a dry snorting sound', he too was sucked below, '& the aeroplane rocked & rolled – miles from anywhere, off Newfoundland, while I slept in Rodmell'. Ten years later Virginia drove past a crashed aeroplane near Gatwick and learned afterwards that three men on board had died. 'But we went on, reminding me of that epitaph in Greek anthology: when I sank, the other ships sailed on.'

The sea echoes over and over again in Woolf's work, with the rhythm of moon-dragged tides. Having finished To the Lighthouse, she entered a dark period, exhausted, fighting for breath; yet out of it she sensed the same vision of presence beyond being that she had seen in Brontë and Melville; something 'frightening & excited in the midst of my profound gloom, depression, boredom, whatever it is: One sees a fin passing far out.' It was a deep, cryptic image, hard to diagnose or discern, as she confessed to her diary a year later, summoning 'my vision of a fin rising on a wide blank sea. No biographer could possibly guess this important fact about my life in the late summer of 1926: yet biographers pretend they know people.'

As a boy on holiday in Dorset, I saw a distant glimpse of dolphins, arcing through the water off Durleston Head, a rocky promontory held out in the grey English Channel. As a girl holidaying in Cornwall, Woolf had seen cetaceans too: one family sailing trip in the summer of 1892 'ended hapily [sic] by seeing the sea pig or porpoise'; her nickname for her sister Vanessa, with whom she was extraordinarily close, was Dolphin. And in *The Waves*, the book that followed *To the Lighthouse*, and which became her most elegiac, internalised work, her vision returned as one character watches a fin turn, 'as one might see the fin of a porpoise on the horizon'.

The sickle-sharp shape seen against the featureless sea – something there and not there – is the emblem of knowing and unknowingness. It is not the real dolphin leaping through the waves, or the curly-tailed, boy-bearing classical beast, or the mortal animal sacrificed and stranded on the sand, but something subtly different: the visible symbol of what lies below, swimming through the writer's mind as a representation of her own otherness. In Woolf's play *Freshwater*, a satire on the bohemian lives of Julia Margaret Cameron and Tennyson on the Isle of Wight, a porpoise appears off the Needles and swallows one of the characters' engagement ring; in *The Years*, 'slow porpoises' appear 'in a sea of oil'; and in a vivid episode in *Orlando*, a porpoise is seen embedded in the frozen Thames alongside shoals of eels and an entire boat and its cargo of apples resting on the river bed with an old woman fruit-seller on its deck as if still alive, 'though a certain blueness hinted the truth'.

Woolf made a sensual connection between the porpoise and her lover. Vita Sackville-West, tall and man-womanish – a kind of Elizabethan buccaneer clad in her brown velvet coat and breeches and strings of pearls and wreathed in the ancestral glamour of her vast house, Knole, where the stags greeted her at the door, and even wandered into the great hall – morphed from she-pirate into a gambolling cetacean for Virginia. It was a dramatic appropriation, dragging the strange into the familiar. Perhaps it was no coincidence that Shakespeare – for whom gender and species were fluid states – often linked whales, living or stranded, with royal princes; or that Woolf's name evoked both the queen and her colony.

At Christmas 1925 the two women, who'd just spent their first night together, went shopping in Sevenoaks, where they saw a porpoise lit up on a fishmonger's slab. Virginia elided that scene with her elusive paramour out of the sixteenth century into the twentieth, Vita standing there in her pink jersey and pearls, next to the marine mammal, both curiosities. 'I like her & being with her, & the splendour,' Woolf admitted to her diary like a schoolgirl, 'she shines . . . with a candlelit radiance, stalking on legs like beech trees, pink glowing, grape clustered, pearl hung . . . so much in full sail on the high tides, where I am coasting down backwaters.' 'Aint it odd how the vision at the Sevenoaks fishmongers has worked itself into my idea of you?' she wrote to Vita two years later, and proceeded to replay the image at the end of *Orlando*, when her gender- and time-defying hero/ine returns home in 1928 – 'A porpoise in a fishmonger's shop attracted far more attention.' Meanwhile Vita made her own boast, of 'having caught such a big silver fish' in Virginia.

Orlando is an updated fairy tale which collapses four centuries of English history into a whimsical modernist fantasy. History rushes by, briefly arrested in close-up, acid-trip details: the grains of the earth, the swelling river, the long still corridor in Orlando's sprawling palace which runs as a conduit into time, as if a production of *The Tempest* were being acted out silently at the end of its wood-panelled tunnel. Orlando is both player and prince, like Elizabeth, or Shakespeare's Fair Youth, Harry Southampton, animal and human, a chimera out of a Jacobean frieze, 'stark naked, brown as a satyr and very beautiful', as Virginia saw Vita. As the deer walked into Knole's great hall, so Orlando moves through species, sex and time; she too might become a porpoise strung with baroque pearls, animating the unknown sea.

There is more than a little of Melville's playfulness in *Orlando*. Woolf read *Moby-Dick* in 1919, in 1922, and again in 1928. In 1921, in her two-paragraph prose poem 'Blue and Green', she wrote a remarkably vivid picture of a part real, part fantastical whale inflected with her recent reading: 'The snub-nosed monster rises to the surface and spouts through his blunt nostrils two columns of water ... Strokes of blue line the black tarpaulin of his hide. Slushing the water through mouth and nostrils he sings, heavy with water, and the blue closes over him dowsing the polished pebbles of his eyes. Thrown upon the beach he lies, blunt, obtuse, shedding dry blue scales.' In *Orlando*, she uses a line from a sea shanty in one of Melville's chapters, 'So good-bye and adieu to you, Ladies of Spain,' and his influence is felt elsewhere in the book, not least when, halfway through a long sentence set in the seventeenth century, Woolf informs her reader of the precise moment at which it was written, 'the first of November 1927', in the same way that Melville time-codes his chapter 'The Fountain' 'down to this blessed minute (fifteen and a quarter minutes past one o'clock P.M. of this sixteenth day of December, A.D. 1850)'. 'For what more terrifying revelation can there be than it is the present moment?' Woolf wrote. 'That we survive the shock at all is only possible because the past shelters us on one side and the future on another.' She even had an active interest in science fiction, prophesying a machine that could connect us with the past, as well as a telephone which could see.

As Orlando leaves the eighteenth century and changes sex, her Shakespearean origins are reflected in the nineteenth-century weather: 'the clouds turned and tumbled, like whales', reminding her, like Keats, 'of dolphins dying in Ionian seas', while the carriages in Park Lane conjure up 'whales of an incredible magnitude'. In these 'unfathomable seas' the natural world takes on an erotic charge. For Ishmael, Melville's unreliable narrator, the shape-shifting sperm whale is freighted with the ocean's obscure desire, while the sea reflects Narcissus examining his own beauty. As the equally unreliable Orlando reaches the eighteen-forties – when both *Moby-Dick* and *Wuthering Heights* took shape – she declares, like Cathy, 'I have found my mate. It is the moor. I am nature's bride.' And in another Melvillean image, she sees a ship sailing through the bracken while her lover recites Shelley and watches it cresting a white wave which, like the Serpentine bridge and the white whale, represents a thousand deaths. (It is telling that Woolf compared George Duckworth, her half-brother and abuser, to 'an unwieldy and turbulent whale'.)

As a writer, Virginia felt she had to conduct a transaction with the spirit of the age; and so she played with the ages – an artist has to stand outside like a seer, or in between, like a medium. Slowed down compared to the speeding time suffered by the humans around her, Orlando lives for four hundred years, as long as an Arctic whale; like Moby Dick, she too seems immortal and ubiquitous – 'for immortality is but ubiquity in time', as Melville says. Perhaps she cannot die. At the end of the book, Orlando drives her fast car back to her country house. Swapping her skirt for whipcord breeches and a leather jacket, she wanders through her estate to a pool which is partly the Serpentine, partly the sea, 'where things dwell in darkness so deep that what they are we scarcely know … all our most violent passions'. It is her version of Gatsby's orgiastic future that recedes before him, beating against the current, borne back ceaselessly into the past, just as Ahab's ship sinks in a sea rolling on as it had done for five thousand years – its biblical age.

In fiction and reality and in between, Woolf continued to delve in a pre-Darwinian, aquatic uncanny. During a visit to Loch Ness in 1938, she met a charming couple at a lochside inn 'who were in touch … with the Monster. They had seen him. He is like several broken telegraph posts and swims at immense speed. He has no head. He is constantly seen.' And in another gothic episode, like one of the gruesome Victorian family stories she liked to retell, Virginia recorded the fate of Winifred Hambro, wife of a wealthy banker, who drowned in the loch when their speedboat burst into flames. Her husband and sons swam to safety, but despite being a good swimmer, and pulling off her skirt, she sank. 'Loch Ness swallowed Mrs Hambro. She was wearing pearls.' It was said that because of its steep sides, the loch never gave up its dead: divers were sent down to recover the body and the pearls, which were worth thirty thousand pounds, but reported only a story of a sinister underwater cave, warm and black. In *To the Lighthouse*, the artist Lily imagines throwing herself off the cliff and drowning while looking for a lost pearl brooch on the beach. The lustrous pearls, lit from within, slide into the abysmal darkness; the sensual product of the sea, they once more become the eyes of the dead.

That same year, 1938, Woolf dwelt on something else elusive. In her essay 'America, Which I Have Never Seen', she looks out over the ocean to a continent she will never visit, sending her 'Imagination' to explore it for her, like Odin's all-seeing ravens or an updated Ariel, 'to fly, | To swim, to dive into the fire, to ride | On the curl'd clouds'.

'Sit still on a rock on the coast of Cornwall,' her spirit-familiar tells her, 'and I will fly to America and tell you what America is like.' Passing fishing boats and steamers, soaring over *Queen Mary* and several aeroplanes, Imagination reports back: 'The sea looks much like any other sea; there is now a shoal of porpoises cutting cart wheels beneath me.' She makes landfall, like an exhausted swallow or monarch butterfly arriving on the Cape. Then the Statue of Liberty looms, and New York: a century on from when Ishmael looked over its harbour, it is now overlooked by 'immensely high towers, each pierced with a million holes'; each a white lighthouse of its own.

This sea is as freighted as any in *Moby-Dick* or *The Tempest*, and it suffuses Woolf's most intensely felt work. *The Waves* is flooded with a stream of consciousness, invoking the voices Virginia heard in her head on the days and nights when her condition drove her to talk unintelligibly for hours on end, and when she lay confined to her bed in her house by the river at Richmond, 'mad, & seeing the sunlight quivering like gold water, on the wall, listening to the voices of the dead'.

The book follows six characters, loosely based on Woolf's siblings and friends, as they move from schooldays to adulthood, drawn together by the loss of their dead friend Percival, just as Virginia lost her brother Thoby. Each section is separated by descriptions of the sea which are more painterly or musical than literary, redolent of Turner's seascapes or Britten's sea interludes. Surging and ebbing, borne on the immemorial Thames and its intimations of the ocean beyond, the water seeps into lives landbound by London. It is the same refined, barbaric world undermined by Conrad's heart of darkness and Eliot's waste land; where the river sweats oil and tar, and barges drift past Greenwich Reach to Gravesend. Bernard, Susan, Rhoda, Neville, Jinny and Louis speak in the voices of the long or recently dead, voices submerged by civilisation and the city-drowned suburbs, crushed by their pressure. They long to escape. They speak as their creator spoke, in perfectly composed sentences which sound like quotations from their own fiction. 'One was compelled to listen even when she only called for more milk,' wrote a friend after spending an evening with Virginia. 'It was strangely like being in a novel.'

'Here, in this room, are the abraded and battered shells cast on the shore,' Jinny says. Louis – for whose voice Woolf drew on Eliot, whom she saw as a slippery eel – is summoned to 'wander to the river, to the narrow streets where there are frequent public houses, and the shadows of ships passing at the end of the street'. Bernard feels the world moving past him like 'the waves of the sea when a steamer moves'; he dreams of going to Tahiti and watching a fin on the horizon. All along the river, time leaks; the Thames is a time machine. We might be standing by Orlando's side as she looks out from her Blackfriars house, past the empty warehouses of empire once filled with colonial plunder and on with the heaving tide, swelling high and brown along the Embankment while twenty-first-century tourists pass by.

This city, through which Virginia wandered in a state of *inner loneliness*, is a place of ritual and sacrifice as much as of trade and progress, coursed with a relentless tide of humanity undone by death, as though it would flow backward out of time. 'It was this sea that flowed up to the mouth of the Thames,' as she wrote in *The Voyage Out*, 'and the Thames washed the roots of London.' And as the characters of *The Waves* watch the steamers sail from the imperial city, they bear witness to the same sea-reach where Marlow saw that 'we live in the flicker, but darkness was here yesterday'.

Under the pavements of Piccadilly – 'the descent into the Tube was like death' – Jinny feels the trains running 'as regularly as the waves of the sea'. Neville – based on Woolf's whinnying intimate, Lytton Strachey – reads a poem, and 'suddenly the waves gape and up shoulders a monster' (an image to be replayed in Iris Murdoch's *The Sea, the Sea*, with its own *Tempest*-inflicted story); he sees his life as a net lifting 'whales – huge leviathans and white jellies, what is amorphous and wandering ... I see to the bottom; the heart – I see to the depths.' And in a passage auguring her author's own fate, Rhoda imagines launching a garland of flowers over a cliff, beyond 'the lights of the herring fleet', to 'sink and settle on the waves', and she with it, like Melville's Billy Budd or Hamlet's Ophelia. 'The sea will drum in my ears. The white petals will be darkened with sea water. They will float for a moment and then sink. Rolling me over the waves will shoulder me under. Everything falls in a tremendous shower, dissolving me.'

This is a fractured, brutal sea, far darker than Orlando's brocaded dreams. It is 'violent and cruel', a new myth enacted under a sky 'as dark as polished whalebone'. Woolf's prose catches that dark and light, washed over by 'water that had been cooled in a thousand glassy hollows of mid-ocean', inundating and uncaring, everything and nothing: an empty eternity, 'no fin breaks the immeasurable waste'. The modernists had found their sea in Melville, an alternative, absolute power with which there is no dialogue, no debate. It stood beyond the depredations of a violent, unequal century, yet was filled with stories like those wreaths cast into its depths, dashed with votive offerings and ghosts of the living and the dead: 'There are figures coming towards us. Are they men or women? They still wear the ambiguous draperies of the flowing tide in which they have been immersed.' The shores are patrolled by phantoms abandoned to the waves. 'It is strange how the dead leap out on us at street corners, or in dreams.'

As we sit around her stove, peering into its glowing interior, Pat tells me she remembers how she has lost good friends, but they came to see her as they died, even though they were far away.

She raises her foot to kick a new chunk of wood into the furnace.

'Do you have visitors like that?' she asks me. 'They do the rounds before they die.'

I see them walking up the beach to her, nonchalantly passing on the news of their passing. I remember walking on another beach with my mother by my side, waiting to see the sun slip behind the shadow of the earth at noon. I remember her greyness at the end, when all colour had been leached out of her, out of her red hair and her high cheekbones, lying there, so still, in her hospital bed. And I wonder if I'll see my loved ones, pacing up the sand or peering round my door as the latch lifts and clicks.

At night, Pat lies in her wooden, boxlike bed, face up. Even in her unconscious she is open to the sea just beyond her window. Does she dream of her dogs, or do her dogs dream of her?

'My dreams got too heavy,' she tells me, as the stove warms my outstretched hands with its iron glow. 'So I stopped having them.'

By the water, terror and beauty go hand in hand. The Pilgrims who came to the Cape – sailing at two miles an hour rather than travelling in the instant accomplished by Woolf's Imagination – found this brave new world a 'naked and barren' place. Soon after *Mayflower* anchored off what would become Provincetown, Dorothy Bradford, wife of their leader, William Bradford, fell overboard and drowned in the harbour.

This naked and barren coast is now studded with lighthouses, symbols of the lost and found. They offer haven and home, safety and hope, welcome and melancholy. They're the land's last markers, measuring out the Cape's confusing asymmetry; you never really know where you are here, no matter how much you might consult the compass or look to the sun. You lose your bearings, find them, then lose them again. But the lighthouses hardly help. They're invested with the strangeness of the sea, only emphasising our separation. For Mrs Ramsay, the lighthouse represents 'our apparitions, the things you know us by'. She looks out 'to meet that stroke of the lighthouse, the long steady stroke ... until she became the thing she looked at'. As she wrote her novel, Virginia saw Vita as a lighthouse, 'fitful, sudden, remote'.

Once fuelled by the whales on whom their beams still fall, these towers of light map out the Cape like one of those illuminated museum displays where you'd press a button to turn on a little wavering bulb. They flash their characteristics, coded signatures like a cetacean's metronomic clicks: from the unseen light at Nauset, which reveals itself only by an anonymous ten-second sweep over the horizon as if it were itself lost, to the six-second signal of the Highland Light, where Thoreau was entertained by the keeper in his 'solitary little ocean-house', and where his bedchamber was flooded with light 'and made it bright as day, so I knew exactly how the Highland Light bore all that night, and I was in no danger of being wrecked'; and on, from Long Point's green dash every four seconds to Wood End's red twitch every ten, looking out over the waters in which in the US submarine S-4 was accidentally hit by a coastguard vessel, sinking one hundred feet to the sandy bottom in the winter of 1927. Wood End Light was no use to it now, down there in the darkness.

Most of the forty-strong crew died – not drowned, but poisoned by bad air like canaries – and storms prevented the rescue of the six survivors trapped in the torpedo room, the only remaining air pocket. A second submarine, S-8, was able to communicate with them by an oscillator attached to the hull of their vessel, transmitting Morse code through the metal skin that separated the men from the world above and the sea around them.

'Is there any gas down there?'

Their officer, twenty-five-year-old Lieutenant Graham Fitch, replied, 'No, but the air is very bad. How long will you be?'

'How many are you?'

'Six. Please hurry.'

Attempts to run fresh air using a hose were frustrated by the high seas. Lieutenant Fitch tapped out a terse entreaty.

'Hurry.'

Then, later,

'Is there any hope?'

'There is hope. Everything possible is being done.'

That night the S-8 began sending out a message, over and over again.

LIEUTENANT FITCH: YOUR WIFE AND
MOTHER CONSTANTLY PRAYING FOR YOU

It took until the following morning, sixty-three hours after the submarine had sunk, for Fitch to tap out his final reply.

'I understand.'

A year later, when the vessel was salvaged, the divers 'found a spectacle that moved them, hardy and inured as they are to horror, to deep emotion. Near the motors, arms clasped tightly about each other in protecting embrace, were two enlisted men, apparently "buddies". The divers tried to send them up thus locked together, but the hatch was not wide enough and they had to be separated.'

If we are lucky, the way we leave this world will become the way we lived in it. According to Sylvia Plath's Lady Lazarus – briefly brought back to life online in her creator's flatly expressive voice, rocked shut like a seashell, unpeeling her skin in the same breath – dying is an art. Plath's fatal, watery dreams merge in her 'Ariel', as her imagination flying over Dartmoor's tors to evoke a distant glimpse, 'a glitter of seas' and a child's cry, an echo of Icarus, the winged boy fallen from the sky.

Plath grew up on the southernmost tip of Boston's North Shore, a site now overflown by planes taking off and landing at the international airport which has turned the streets she knew into potential runways. 'My childhood landscape was not land but the end of land,' she recalled, in the last piece of prose she ever wrote, now living in London, '– the cold, salt, running hills of the Atlantic. I sometimes think my vision of the sea is the clearest thing I own. I pick it up, exile that I am, like the purple "lucky stones" I used to collect with a white ring all the way round, or the shell of a blue mussel with its rainbowy angel's fingernail interior; and in one wash of memory the colours deepen and gleam, the early world draws breath.' She remembered other debris, too, 'sea-beaten' nuggets of brown and green glass – 'blue and red ones rare: the lanterns of shattered ships?' And she recalled her mother's stories of wrecks picked over by townspeople on the shore like an open market, 'but never, that she could remember, a drowned sailor'.

As a child, Sylvia had crawled into the water – 'Would my infant gills have taken over, the salt in my blood?' – and later taught herself to swim. She believed in mermaids more than in God, and saw the sea as 'some huge, radiant animal'. At six years old, she and her younger brother watched a hurricane sweep over Cape Cod Bay, 'a monstrous specialty, a leviathan. Our world might be eaten, blown to bits. We wanted to be in it.' The storm rocked their house, howling outside the black window in which her face was reflected like a moth trying to get in. The sea was her past, sealed off like a ship or a god in a bottle, 'beautiful, inaccessible, obsolete, a fine, white flying myth'. But it was her now, too, and it seethed through her poetry – 'A far sea moves in my ear' – sucking her body of blood till she was as white as a pearl.

As a teenager Sylvia returned to Cape Cod, which she had known all her life. She was the same age as Pat, swimming and sunbathing, modelling herself on Marilyn Monroe with her bright white bikini, her beach tan and her sun-bleached hair. She worked here in the summer as a mother's helper, looking out longingly from her duties to the 'blue salt ocean'. It promised all the ecstasy of expectation. But it was disturbing, too. Plath was already experiencing deep bouts of doubt and depression. In the summer of 'fifty-two she read *To the Lighthouse*, underlining the lines 'Of such moments ... the thing is made that remains for ever after'; in her beach bag was a copy of *Orlando*. The following summer of 'fifty-three, she swam out into the sea, and kept on swimming, trying to emulate her heroine, 'Only I couldn't drown.' Two days later, and two days after that, she attempted again, and again, to take her own life.

Her mother tried to send Sylvia to a friend in Provincetown to recover. This place might have healed her. Instead she was admitted to an asylum where she felt she had been put under a glass bell. Electrodes were attached to her temples and dials turned on a console that resembled the equipment my father used to test the resistance of electrical cables. Instead of testifying to her resistance, they systematically attacked her memory, the most precious thing she possessed: the means of her imagination.

In 1957 Plath came back to Cape Cod with Ted Hughes. The newly-weds spent a seven-week summer here, a delayed honeymoon. It was a heady time. They stayed in a cottage at Eastham, down the coast from Provincetown. Sylvia introduced her lover to the beaches. Her favourite, at Nauset, named after the original inhabitants, was a bike ride away, and in her journal she recorded the rhythm of their lives, 'write, read, swim, sun', as if it would never end.

Walking between Nauset and Coast Guard Lights, the shore where the Pilgrims first sighted land, the pair found a sandbar, shallow and smooth. Floating with her hands and feet bobbing like corks, her wet hair trailing as if to trap the fish, Plath felt a sense of power and glory. The possibility of a new life lay ahead; the elemental flux refixed her; she was recalibrated to the ocean. The pair were caught in the 'great salt tides of the Atlantic'; for Plath, marriage had set 'the sea of my life steady'. To enshrine it, the two poets recorded the shore. She saw fiddler crabs in a dried pool at low tide, scuttling to dig themselves into the sand with their one gigantic claw; they were denizens of a 'weird, other world'. He would recall the 'pre-Adamite horse-shoe crabs in the shallows', their 'honey-pale carapaces' and the 'wild, original greenery of America'. They were post-war, modern people; they could fly over the ocean that had defied Shakespeare, Keats and Woolf. Here they found the oldness of a new world that Hughes had imagined before he'd ever seen it, defined by the ocean and 'the whaled monstered sea-bottom'. His words echoed Eliot's, who'd sailed here as a young man and found 'hints of earlier and other creation: | The starfish, the horseshoe crab, the whale's backbone'.

While her husband thought of the men who drowned out there, 'Where darkness on Time | Begets pearl, monster and anemone', Sylvia was reading Woolf again. She finished *The Waves*, disturbed by its 'endless sun, waves, birds, and strange unevenness'. She felt Woolf's writing made hers possible: 'I shall go better than she.' The crystalline memory of that time would be set in the shadows of the future. Hidden in the sardonic snarls of 'Daddy' – partly written about her father, whom Plath saw as a dark Prospero, and partly about her partner, who, like everyone else, would leave her in the end – is a sudden shining invocation of happiness: the sight of a seal, 'a head in the freakish Atlantic | Where it pours bean green over blue | In the waters off beautiful Nauset'. These lines would come to haunt Hughes, who, like Plath, believed in signs and wonders, and who, years later, recalled that same shore descried in Sylvia's 'seer's vision-stone' – one of the white-ringed purple stones she picked up from the beach. Through it he saw her brown shoulders, now in a black bathing suit, secure on her childhood's shore. On the beach, everyone is immortal.

But Plath knew none of this was real. She was engulfed by the tyrannical beauty of this place. 'I'd rather not live in this gift luxury of the Cape, with the beach & the sun always calling ... I need to end this horror: the horror of being talented and having no recent work I'm proud of, or even have to show.' The world was speeding past. As they trudged under the hot sun along the sandy roadside of Route 6, where the pines grew short and looked as young and as old as the country itself, 'deathly pink, yellow and pistachio colour-ed cars' shot by, 'killer instruments from the mechanical tempo of another planet'. The pair were relieved to reach the beach, where the red-and-white Nauset Light surveyed waves five miles long.

Plath left the sea for Smith College in Massachusetts, where she was taught Melville by Newton Arvin, known as 'the Scarlet Professor' (and was also lover to Truman Capote). She lived with Hughes in the nearby town; someone in Provincetown who'd been to the same college told me that his landlady had complained about being unable to get rid of the smell of perfume which her former tenant, a poet, had spilled on her desk. Plath was rereading *Moby-Dick* and finding solace in it, 'whelmed and wondrous at the swimming Biblical & craggy Shakespearean cadences, the rich & lustrous & fragrant recreation of spermaceti, ambergris – miracle, marvel, the ton-thunderous leviathan'. She even imagined herself, 'safe, coward I am', aboard a whale ship 'through the process of turning a monster to light and heat'.

Four years later in her London flat, in the winter of 1962 – the coldest for two hundred years, when the capital lay covered in snow and ice and she was reduced to wrapping herself in her coat and walking to the telephone box to plead with her husband to come back or to tell him to leave the country forever – Plath wrote a radio talk for the BBC. It was entitled 'Ocean 1212-w', after her grandmother's telephone number on Boston's North Shore. In it, she recalled the hurricane which had turned a childhood afternoon sulphurous and dark. 'My final memory of the sea is of violence,' she wrote, '– a still, yellow day in 1939, the sea molten, steely-slick, heaving at its leash like a broody animal, evil violence in its eye.'

She would never record the broadcast. Two days later, with the snow still lying on the city's streets and her children in their beds, she carefully sealed the kitchen with wet towels, turned on the gas, and put her head in the oven.

On the winter shore, where the sea rises higher than my head, Mary and I scramble down a great sandy scree. All around us purple and green stones stand proud where the air has scoured around them, leaving them balanced on gritty pedestals like miniature megaliths. Others have tumbled down the cliff, leaving a drunken trail behind them. We wonder if they continue to roll while we're not looking, only stopping when we stare at them. Clumps of compass grass, true to their name, draw circles around themselves as they whip about. Their stalks are brown and dead, all colour drained away by the dunes. But they've been resurrected by the wind.

We associate the beach with life, with warmth and the sun; not this numbing down, this low-season, reiterated death. At Race Point, the old lifesaving station stands empty; there'd be no handsome New Englander ready to throw me a lifebuoy if the ocean or my heart decided this was my last swim. The *New York Times* of 23 August 1883 recommends this shore to 'those looking for good surf-bathing', so empty even on a summer's day that 'if one wishes to go in in his bare pelt he may do so without the fear of shocking propriety'. Although it is January and the air temperature is about to fall to minus twenty, I duly heed the *Times*'s advice.

You have to take your chances with this ocean, roaring, rising high, utterly elemental and unharnessed, more like a mountain range than water. I wait for my moment, trying to judge when to get in, if I should get in at all. The water is warm compared to the air. And just as the waves roll me around, like stones in a barrel, I feel abraded by the winter, made raw, my bones exposed by the cold, both withdrawn and peeled back by it, as if I took off my skin too when I skinny dip. The cold removes sensation then restores it. The hardness of winter is an absolute in an equivocal world.

Each day seems colder than the last. The black ducks and eiders come into shore, seeking the respite of the sand. But the buffleheads, the smallest ducks – so sharply marked in stark black and white that they look more like feathered orca than the buffalo they supposedly resemble – stay out, diving fast and brave beneath the surface. It is strange, this comfort birds offer us, considering the discomfort they seem to suffer. Polar animals occasionally arrive here: a beluga whale once nosed around the moored boats, and a bowhead whale has appeared in recent years, caught up with its right whale cousins, its slow smooth bulk bringing an Arctic chill to the bay.

The North does not feel far away: in Pat's paintings of frozen seas, echoes of the icy waters in which her great-grandparents perished, or in the thick snow through which I wade to get into the water after driving back from New Bedford with Dennis in a blizzard whose swirling blinding flurries rouse like ice monsters, turning the Cape landscape into the Russian steppes. A few doors down from here is the home of the polar explorer Donald MacMillan. His wooden house stands over the beach as if it had arrived here on a floe, stranded like all those fatal ships – *Resolute*, *Erebus*, *Endurance*. Surrounded by a white paling fence, its lawn could have been a corral for the polar bear whose stuffed and yellowing remains rear over visitors to the town museum and in whose skin MacMillan might be hidden, like some Inuit shaman.

At MacMillan Wharf, named in his honour, the trawler *Tom Slaughter* out of Gloucester has come into the harbour, seeking shelter; the young bearded crew might have been fishing here for a hundred years. Dovekies – little auks, their wings whirring like clockwork toys, wound up by their economic binomial, *Alle alle* – fly in, too. Named after the Swedish diminutive for dove, they're also known as ice birds. Out in the open ocean they assemble in their millions, flying through the water on their wings to feed. They're seldom seen this close to shore, these brave, stubby birds; some storm far out at sea has driven them in, along with all the others. In Charles Kingsley's *The Water-Babies*, young Tom, become amphibious, goes north in search of Mother Cary, passing tens of thousands of birds, 'blackening all the air; swans and brant geese, harlequins and eiders, harelds and garganets, smews and goose-anders, divers and loons, grebes and dovekies, auks and razorbills, gannets and petrels, skuas and terns, with gulls beyond all naming and numbering', ending up in the frozen waters 'where the good whales lay, the happy sleepy beasts, still upon the oily sea'.

Everything has been slowed down; a slow pace for a slow place. A Provincetowner lends me Sten Nadolny's *The Discovery of Slowness*, a retelling of the life of Sir John Franklin, would-be discoverer of the North-West Passage. Nadolny's prose itself slows the reader as he lays out Franklin's story: his slowness to respond to his teachers in his Lincolnshire school; his slowness to draw fire on the enemy at the Battle of Copenhagen, where Nelson appears, never quite appearing; his slowness to stop strangling a Danish soldier with his bare hands; his slowness to respond to a man overboard when sailing with his uncle, Matthew Flinders, to Terra Australis, where Franklin would become lieutenant-governor of Van Diemen's Land; the way he leaves a pause before he answers, the consideration he gives to what he is about to say, the manner in which he seems to militate against an accelerating century. All these happen at a glacial pace. Slowly moving through history, slowly moving across the ice, eyes protected against snow blindness by wooden spectacles, the sea slows him down too, as well as providing him with a kind of certainty. 'As long as there was the sea, the world was not wretched.'

But most of all, the explorer's slowness slows him to his ending on *Erebus* – named after the god of darkness – out on the slow-moving, ever-reforming ice which, it seemed then, would never end, and where he succumbs to a stroke, leaving his starving men to eat each other. To the native hunters who saw them, the lost, bedraggled adventurers were dead men walking.

In the dark morning I pace the deck, waiting for the light and the tide to arrive. It is shocking when the sun comes up, lifting over the sea, as if I'd forgotten it was there.

As I swim off Pat's bulwark, there's a commotion on the sea wall. Running along its concrete edge in front of the Icehouse, a fish storage block turned into apartments, are a pair of foxes. They're in peak condition, red and ginger and grey, brushes held out proudly as though they'd come straight from the hairdresser's. 'Would you like some product on that?' They follow one another nose to tail, resplendent in their ordinary otherness, calmly trotting over the deck before vanishing down the alley.

Pat calls me in after swimming to warm myself by the stove. I crouch over it, dripping, clad in a towel and my trunks. Its cast iron is piled with bricks to radiate its heat, creating a votive altar on which little metal figures perch as household gods: a long-eared African running dog; a bronze rat, wired to a whale vertebra; and smallest of all, a tiny cold-painted pale-green monkey with wire-thin curling tail. The stacking and burning of wood from the cords piled outside; the lichen-covered logs scattered with frozen cats' turds; the loading into the barrow and the trundling to Pat's door (Caliban would complain that there's wood enough within) – all this proposes a primal exchange: trees for heat; my body for the sea.

Pat stokes the fire with a poker, the incandescence catching the light in her eyes and her hair. Lulled by the dumb cold outside and the dry warmth within, we talk about her past: how her mother left her father when Pat was three years old; how her uncle, a naval commander, arrived on his aircraft carrier in New York harbour, and how she climbed right up the rigging. She still remembers that feat: the fourteen-year-old hoyden caring nothing for convention, running up the ropes as if to spy out her future over Manhattan and the sea beyond. She calls her father Daddy and repeats his full name, Ernald Wilbraham Arthur Richardson, as if to draw him near.

All too soon the afternoon darkens into night and the windows, which briefly burst into life with a final flare of the setting sun, turn into black screens. 'What a fine frosty night,' says Ishmael; 'how Orion glitters; what northern lights! Let them talk of their oriental summer climes of everlasting conservatories; give me the privilege of making my own summer with my own coals.'

I retreat upstairs to my eyrie. It's like living in the nineteenth century: the creaking wood, the clicking latches, the fugitive light; the smell of mothballs and departed summers; reading into long nights on the salvaged sofa, my legs covered by an 'Early's Settler' blanket labelled with an embroidered Red Indian head, its thick cream wool pelt like Ernald Wilbraham Arthur Richardson's snow coat; the drowsiness induced by the cold, the lull of a doze; rising in the dark morning as if nothing had changed while I was asleep, to read under the worn bedclothes, a knife beside me to slit the uncut pages of *Ariel*, André Maurois's nineteen-twenties biography of Shelley, as though even old books are still dangerous. And as I read, I'm held under the heavy blankets and quilts, bedded down and layered against the sea and the wind like Ishmael in bed with Queequeg the noble savage, rejoicing in the contrast between the warmth of his nest and the frozen tip of his nose, 'the more so, I say because truly to enjoy bodily warmth, some small part of you must be cold, for there is no quality in this world that is not what it is merely by contrast'. I am most comfortable in the cold; and I seem to be cold so much of the time, whether in the water or wrapped in wool in my attic.

The old-newness of this house, which is actually younger than me, makes it seem as if all these things have always been here, or that they have only just been found: the slowness, the isolation, the layers of clothes worn over and again, too cold to wash, too cold to get dirty; the elemental demands, the forgetting and the remembering, the short days and long nights; the sticky porridge I stir on the stove; the heavy red kettle I boil for my tea; the dull roar of the antiquated heating system rumbling into action like a ship's engines, as though there were a great steam pump somewhere in the bilges of the house; the clatter of Pat's yard-sale ironware crockery, with its transfer prints of birds and English idylls recreating the unremembered land that she lost; the pitted cutlery and its cracked bone handles; the faded rugs, the wonky furniture, the globular lights with their fitful glow, the tattered aristocracy and resilient fabrics of the past: stone, iron, sand, beach.

I think of all the people who have stayed in Pat's house, living over its fierce mistress. She has to remember to forget. How else could she deal with all the visitors who come and go? The night is still, still cold. Snow turns to ice on the sand, spatters my deck like a Greenland whale ship. Ducks talk in the darkness. The sea might be full of extinct animals out there. The cloud cover bounces residual light off the snow.

Early on a bright Sunday morning, Dennis and Dory and I go clamming off the breakwater. On one side lies the open bay; on the other, frozen moors and marshes. There's such a width of seeing here. The snowy flats lie at low tide, scattered with townsfolk out to gather the harvest of quahogs. From a distance, their little black figures spread across the whiteness, holding rakes and wire baskets, resemble a painting by Brueghel. They're foraging with the other animals. Gulls drop clams on rocks to crack them open; Thoreau recorded that ducks feeding on clams were caught by the shellfish, their feet 'tightly shut in a quahog's shell'. Dunlin stitch through the sand for smaller prey, so intent that they ignore my shadow falling over them.

I tug the wide rake through the sand, feeling for the chink of shell against its tines. The tide is returning like a sheet over a table, and with it the end of the morning's foraging. Dennis carefully replants the clams too small for our basket, telling them we'll see them next year. That night, their elders open up in the pot, each 'little brother' receiving an apology from Dennis as he drops them in. The softness out of the hardness makes me queasy, as if we're doing something we shouldn't. I think of the giant oysters found off the Azores, five hundred years old. We're eating centuries encased in islands; consuming the ocean bed, one pebble at a time.

Still the weather draws in, reducing degree by degree. The crust of the frozen sand crunches underfoot like compacted snow. It's easier to walk the beach in the cold, says Pat. I watch her figure diminishing up the beach, towards the town limits, with her own animal momentum, stomping as if she'll never stop. The brent geese have arrived on the low tide, a gently honking fleet, so familiar from my home shore that I wonder if they've followed me here. These are their pale-bellied cousins, and they look even more pristine against the beach. They forage in the iced-up eel grass. In the dull light their subfusc markings, brown and black and white, seem to distil the land and the sea, connecting them. They gather in a shallow pool, upending their rears.

Overnight the mercury falls even further. I read of lifesavers on the ocean beaches caught in snowstorms that froze their eyes shut. At dawn, the bay is blue with cold. Slabs of light lift and shift over the half-fluid, half-airy surface. It's difficult to tell what is water and what is sky.

Then something extraordinary happens.

Far out, the sea's relative warmth meeting the cold air causes localised clouds to rise, lifting in tendrils from the surface and drifting steadily across the bay. Imperceptibly, they begin to co-alesce, drawing together.

Slowly, I realise I am seeing something miraculous, some phenomenon out of the ages. A natural philosopher ought to be at my side to witness it. But I don't really want it explained. I can't stop watching.

Raising my binoculars I see it all as if framed in one of Pat's paintings. Or is it the other way around? With every second, every-thing changes. The geese lift off into the sky, sharp black shapes against the white-grey. As they fly across my field of vision, the vapour rolls below them, driven by the light. Layers of mist move one way; others in the opposite direction. Dissolving and re-forming, resolving and dissipating, they swirl and dance like the mist on an English river on an autumn morning.

They might be airy icebergs, or the spoutings of spectral whales, or the ragged sails of a ghostly armada. They seem to echo the currents below. Then behind the white wisps even more atten-uated plumes rise and twist like water spouts. There's so much smoke on the water it looks as though the sea is on fire.

It is the single most astonishing effect of weather I've ever seen, as if someone had dipped a brush in the water and drawn it up in swift strokes. The horizon has become one long mountain range of white peaks. A flurry of snow drifts through the air, light and granu-lar, whipped up by a god of water and ice. And just when I think the whole performance cannot get any more beautiful the sun breaks through, throwing the scene into silver relief, sending it all into shards of glitter. The scrims of mist which looked two-dimensional a moment ago are suddenly thrown into three dimensions by the sun's rays breaking through the clouds like the eye of God.

Then the whole thing starts all over again.

Landing on the flattened beach, I follow the footprints I left there last night. The receding tide has left a thousand rivulets in the sand, a venous branching pattern running from single shells or stones, each expanding into miniature deltas which might be the world photographed from twenty miles above. On this shore, leviathans have also lain. John Waters, who has no interest in whales, gives me an old postcard, showing a fin whale shouldering up the sand, with onlookers gathered around like spectators at a seventeenth-century stranding in the Netherlands. Back in old Europe, such a spectacle would have been taken as an omen. But their modern regard is curious and interested, unsurprised. It might be a carnival, a ritual that happens every year, without anyone actually deciding it should.

All whales bear their own stories. Stormy Mayo, Charlie's son, remembers this one well. It was the winter of 1959, and he was still at high school. The whale first washed up on the Mayos' beach, still alive. At high tide the Coast Guard tried to haul it out to sea, 'but the line parted about two hundred yards off our beach ... you see the line still around the tail'. Out in the water the whale began thrashing in its death throes, before coming ashore, down towards the town, not far from where Dennis and I found the dolphin. The artist Elspeth Vevers, who lived across the street, remembers its arrival too. She and her husband Tony, who later painted the scene, tended the whale, keeping it wet with towels. Its jaw was askew, Elspeth recalls, and they gently removed the bricks which people

had thrown into its mouth. All the while it looked directly at them, following them with its eyes.

Then a team of scientists from Woods Hole appeared to take specimens and observations. They found it was a male, weighing forty tons. His body temperature was ninety degrees Fahrenheit; his fins and tail just fifty degrees. It was cold there, on the December beach, but the life was still in him. They recorded his heart beating twenty-five times per minute, one-third the rate of a human; one reason why fin whales can live to one hundred and forty years old, perhaps even older.

But not this whale. Sometime that afternoon, after further attempts to tow him back to the sea had failed, and having given up his data for science, his heart stopped and he took his last little sleep on the shore.

The nor'westerly has returned. It starts gently enough, spooling and eddying over the sea's surface as though it were a cup of tea and someone were blowing on it to cool it down. Then hammered latches rattle. Frames push against panes. Now there are poltergeists knocking at the windows and doors, all those artists trying drunkenly to get back in. I feel I'm in the way of big things: whales, winds, waves. If the house should lean either way, the whole thing might fall apart, the cedar shingles flying off, nails springing out in a hail of tacks, the wind lifting the gables like the flap of a tent.

Forty-eight hours later the front changes direction. Now it's blowing with full force from the south, straight from the bay.

In the darkness of my cabin, the fever increases. Everything is shifting. Something buzzes like a trapped fly. The house has become a theme ride. Or perhaps some spirit has crept out of the water and into the crawl space and is busy shaking the timbers in revenge for some long-forgotten wrong. The wind rattles the tin cowling on the chimney like a destitute begging for money.

My body is gently trembling against the couch as I sit on it. I'm not sure if it's my heart beating, beating for all the cold I have

inflicted on it in the ocean. I feel like a sailor swinging in his hammock, swaying sickeningly to and fro. The noises increase, in an
ever-stranger symphony of dissent. In *The Perfect Storm* Sebastian
Junger describes how fishermen judge the intensity of a storm
from the sounds made by their wire stays and outrigger cables. If
they scream, the wind is Force 9 on the Beaufort Scale; a shriek
means Force 10. Force 11 is a moan. Anything over that is 'something fishermen don't want to hear'.

It is the house shaking, like a washing machine. Perhaps this is
what an earthquake feels like. I look out the windows; I can't open
the door. The sea is crashing at the feet of the house. Suddenly,
stirring me out of what may or may not be my dreams, Pat appears
at my door in the shadows, her hair electrified, silver like my mother's. It's the middle of the night, but she has come to make sure
I'm awake to witness the storm, knowing that it is more important
than sleeping. Even she seems nervous, although her house has
stood worse assaults. I try to go back to sleep, but my mind starts
creating escape procedures, working out what I will save before it
all becomes wreckage, so much flotsam and jetsam.

At dawn, with the arrival of the light, the world has gone quiet.
Sometime during the night, Pat's raft has been torn from the chain
that anchored it to the sea bed. I wade out to retrieve what is left
of it, ragged with torn wood and bent screws, lulling in the what-
did-I-do tide. The beach has been lowered so much that it's now
too far to jump down from the bulwark. Bits of stairs and uprooted
buoys litter the shore. This town is remade after each new storm.
Its heart is jumpstarted by the climate. I am in no danger of being
wrecked.

I think of the people who have climbed this same ladder to my
bed; perhaps they might be climbing up to join me in this animal
loft. Pat tells me that a woman who stayed in the ground-floor
studio – which I can see through the cracks in between the floorboards – complained of the noise made by men having sex up here.

That night in the calm darkness I watch a comet, a blue-green
blur on the edge of Orion, an ancient smudge forty-four million

miles away, glowing with its own gases. To say 'watch' isn't quite right. I only witness it; or rather, it is shown to me by the sky. Even through my binoculars it looks like nothing more than a puff of smoke, the ghost of an exploded star, or a blow from Cetus, the whale that surfs the winter horizon. Nebulous, more cloud than star, its celestial distinction is its indistinction. I can't see it directly, only out of the corner of my vision, as if its magnitude or its meaning were too great to take in. Such visits have long been seen as omens; as Nathaniel Philbrick notes, it was a blue-green comet that appeared in 1618 which prompted the Pilgrims to contemplate their flight from Europe to this coast.

A comet's tail always streaks away from the sun, fluttering like a sail on the solar wind. It may stretch for hundreds of millions of miles: the great comet of 1843 reached halfway across the solar system. This heavenly body is both hot and cold, moving so fast and so slow; even its sea-like colour evokes the paradox of a ball of burning dust and ice, and the possibility that we owe our oceans to water which first arrived on comets or asteroids. All the while it emits x-rays, as if examining the other bodies which it passes.

I know that my friend Mark is looking up at the same phenomenon from southern England. I feel triangulated by the fact, fixed across the ocean by a natural satellite on its vast elliptical orbit; perhaps some flock of astro-geese or school of celestial cetaceans might use it to navigate, migrating from one galaxy to another. Forever roaming, never at home, it slowly swings out to the dark recesses of our system, passing Prospero and Ariel, returning to the sun a thousand or a million years later, only to leave again, a lonely body, like a trans-species whale that will never meet a mate.

Up there is the scary otherworld with its mask off, the infinity into which we are all falling. The green lights flash across the bay. I'm sailing on an uncharted sea. Starry animals dance across the sky, born out of the water. My heavens, I think: sea or sky, it doesn't seem to matter which. Everything is the same.

I'm not even sure which way is up any more.

SOMETHINGAMAZING

In March 1845, Henry David Thoreau left the New England town of Concord and set out for the woods around Walden Pond and, having borrowed an axe for the purpose, began to cut down white pines on land owned by his friend Ralph Waldo Emerson. As Thoreau shaped the timbers, he stopped to eat his lunch of bread and butter, reading the newspaper in which it was wrapped. He was constructing a hut, an unfixed locality, citing Indian teepees as eminently suitable to their purpose and eschewing the mortgages to which his fellow Concordians were shackled. The philosopher hoped for a reconnection, albeit a temporary one. 'We no longer camp as for a night,' he wrote, 'but have settled down on earth and forgotten heaven.' He sought utopia by the water's edge, just as William Blake, after 'three years' slumber on the banks of the Ocean' in his Sussex cottage, had proposed a new Jerusalem.

The son of a pencil-maker, Thoreau was educated at Harvard University, and had briefly been a teacher – he gave it up after finding corporal punishment an offence to his sensibilities. In 1842 his twenty-six-year-old brother had died in his arms, from lockjaw caught after cutting himself shaving. Thoreau was consumed with grief. He later proposed marriage to a young woman, and was rejected. Thereafter he would work as a surveyor, measuring out the land while pondering its metaphysics. Now he had retreated to a New England wood (the same wood he and his friend Edward Hoar had managed to set on fire when camping there the previous year). Even in his attempt to distil utopia into a commune of only one, Thoreau could not help but be part of the greater world. When he bought the frame of his shanty from an Irish navvy – for four dollars and a quarter – there was another price to be paid: by the family effectively evicted by the transaction, and by their cat, which went feral and was killed when it trod in a trap laid for woodchucks.

By the sandy shores of Walden, in among its trees, Thoreau unpacked his shack and trundled it, bit by bit, by cart to his pondside site, laying out the parts to dry and bleach in the sun (having been told by 'a young Patrick' that another Irishman had already stolen all the nails). He excavated a cellar underneath the hut where he could overwinter his potatoes; all houses, he wrote, were 'but a sort of porch at the entrance of a burrow'. He was digging into the land like a badger.

Friends helped him raise the frame of his house, in the way Shaker barns were raised as communal efforts, and he took possession on the Fourth of July. Construction continued as he built his chimney using stones from the pond, claiming them, as he did the timber, under squatter's rights. Thoreau relished labour for its own sake. If we all built our own houses, he said, 'the poetic faculty would be universally developed, as birds universally sing when they are so engaged'. Instead, 'we do like the cowbirds and cuckoos, which lay their eggs in nests which other birds have built'.

Thoreau saw transcendence in everything he did and wove philosophy out of the commonplace. He lived in his own quietude, finding it more fruitful than the alternative. 'The man I meet,' he said, 'is seldom so instructive as the silence he breaks.' Made insular by the beauty of the natural world, he wrote about himself, reasoning that 'I should not talk so much about myself if there were any body else whom I knew as well.' And if he lived in a dream, what of it? What else should we do with our hours of wakefulness? Nothing more useful than the hours we spend dreaming.

Shingled and plastered, ten foot wide by fifteen foot long, with eight-foot-high posts, Thoreau's hut contained an attic and a closet, a tiny cellar with two trapdoors, and a brick fireplace. Outside was a woodshed. All built by himself, for $28.12½, including $1.40 transportation ('I carried a good part on my back'). He reckoned the price to be less than the annual rent paid by most of his townsfellows, another good reason to recommend the effort. 'I brag for humanity rather than for myself.'

Walden Pond is wonderfully cold on a late spring afternoon. I push out from shore, reluctant to go far, knowing Thoreau's plumbline drew more than one hundred feet as he surveyed its dark extent. In midsummer this place is alive with people and noise, with children and picnics and canoes. Today, it is silent and still, save for the concentric rings sent out by my body. It's hard to believe that Boston is only half an hour away. The railway runs close by; it did so in Thoreau's day, although the whistling trains merely reminded him of his solitude. On the other side of the road there's a replica of his hut, complete with a bunk covered with a green woollen blanket; a single bed is as eloquent as any obituary, or any passing star. 'Why should I feel lonely? is not our planet in the Milky Way?' reasoned its sole occupant. 'We are wont to imagine rare and delectable places in some remote and more celestial corner of the system ... I discovered that my house actually had its site in such a withdrawn, but forever new and unprofaned, part of the universe.'

Thoreau's daily swims were a communion, undertaken at dawn. 'I got up early and bathed in the pond; that was a religious exercise, and one of the best things I did ... Morning brings back the heroic ages.' Ever analytical, he tried to examine the very colour of the water – 'lying between the earth and the heavens, it partakes of the color of both' – and noted that a glassful held up to the light was crystal clear. And in a dreamlike image, he observed that when the body of a bather – presumably his own – was seen through the water, it was 'of an alabaster whiteness', the limbs 'magnified and distorted withal'. Nowadays analysis of the pond water in summer betrays a high volume of human urine.

Walden Pond became an extension of Thoreau's hut; he even used its white sand to scrub his floor, wetting and scattering it before brushing it clean. Were all these activities, so exactingly and intensely enumerated, ways of forestalling darker thoughts? The black water offered liberation. 'After hoeing, or perhaps reading and writing, in the forenoon, I usually bathed again in the pond, swimming across one of its coves for a stint, and washed the dust of labor from my person, or smoothed out the last wrinkle which study had made, and for the afternoon was absolutely free.'

In nearby Concord, after swimming in the clear green weedy river behind Nathaniel Hawthorne's house, hoping he wouldn't mind, I visit the basement of the town library, where the curator emerges from the vaults with an armful of oversized documents protected by plastic sleeves. They flap rather worryingly in her hands as she lays them out on the table with a mixture of reverence and familiarity. I am allowed to inspect, but not photograph these relics, prophylactically sheathed in mylar. The air has been sucked out of them; the human has been excluded. Seen through thin plastic they remain remote, even though they're lying in front of me.

Thoreau's fine draughtsmanship charts the pond as carefully as a navigator might chart the sea; as his plumbline drew water, so Thoreau drew the pond. Tiny numerals record the depths – 30′, 91′, 121′ – and ruled lines divide its expanse in precise, faint ink, as though he had trailed a sepia fishing rod behind him as he rowed across the pond. Thoreau may have sought to disprove those who believed Walden to be bottomless, but in the figures an implicit poetry seeps out; mere mathematics could not confine his thoughts. These points could be constellations as much as charts of a body of water. To him, 'there was no such thing as size', his friend Emerson wrote. 'The pond was a small ocean; the Atlantic, a large Walden Pond. He referred every minute fact to cosmical laws.'

Walden was a microcosm of everything. 'It is earth's eye,' Thoreau said, 'looking into which the beholder measures the depth of his own nature.' Yet in imposing his lines on the wilderness, wasn't he destroying what he observed, even as he'd managed to burn three hundred acres of it by mistake?

In a tall glass case against a pillar that supports the silent mass of books upstairs – who reads them now? Certainly not the people sitting at the tables, their faces turned pale by their laptops – Thoreau's instruments are displayed. The polished wooden tripod and theodolite mark the measure of the man, as if he'd just stepped away to take account of the land around, taking in all that beauty and not believing it.

In 1862, as he lay in a Concord attic, dying of the disease that consumed him at the age of forty-four, Thoreau edited his final text, 'Walking'. It's here, on the table. I copy the words from his own hand, closing the gap between his then and my now. It is his last will and testament, by default: 'I wish ~~this evening~~' – the two words excised in the edit for posterity's sake – 'to speak a word for Nature, for absolute freedom and wildness, as contrasted with a freedom and a culture merely civil, – to regard man as an inhabitant or a part and parcel of Nature, rather than a member of society.'

New England lay empty and full of history, settled and un-settling; it allowed such transformations for its utopians and its Transcendentalists. Emerson had experienced a panoptic epiphany on Boston Common: 'Standing on the bare ground, – my head bathed by the blithe air and uplifted into infinite space, – all mean egotism vanishes. I become a transparent eyeball; I am nothing; I see all.' Other peers of Thoreau took matters to an extreme. Samuel Larned was the son of a Providence merchant who had become obsessed with his own consumption. He'd spent a year living only on crackers; the next year devoted to eating only apples. When he stayed at Brook Farm he declined to drink milk and wore vegan shoes. He also swore at everyone he met: 'Good-morning, damn you.' We might diagnose Tourette's syndrome, but Larned believed that profanities uttered in a pure spirit 'could be redeemed from vulgarity'.

Larned and his young friends were the Apostles of the Newness, marked out by their 'long hair, Byronic collars, flowing ties, and eccentric habits and manners'. In 1842, three of them set off on a walking tour, wearing broad-brimmed hats, sack coats and as-yet unconventional beards. They took no money with them, relying on people they met to provide them with food, although, as Richard Francis comments, 'given their severe dietary restrictions, that wasn't asking much'. When they arrived at Emerson's house, he quickly moved them round to the back door where their swearing wouldn't disturb the neighbours. Their appearance was becoming more outrageous: by the time Larned and his friends came visiting the following year, they were 'peculiarly costumed' in smocks belted about the waist and made of 'gay-coloured chintz'.

The sight of these visionary young men dressed in what looked like blouses got up from flowery curtains must have made a certain impression in the streets of Concord and Boston, as similarly dressed young men would do in Woodstock the following century. Even at the time they were referred to as 'ultra', as if set beyond the pale. That year, 1843, Larned found his natural home in the near-by commune of Fruitlands, whose members declined to enslave animals to plough their fields and used, ate or wore nothing animal

"Fruitlands", The Home of the Alcotts, Harvard, Mass.

or produced by slavery, from cotton to whale oil. Some went naked under the New England sun, while their leader, Bronson Alcott, bathed in cold water, rubbing his body afterwards with a 'friction brush'. He also claimed to be able to enter a kind of altered state, with sparks flying from his skin and flames shooting from his fingertips, 'which seemed erect and blazing with phosphoric light', as if he were a Blakean being. When he closed his eyes, they 'shot sparkles', and in his ears he heard a melody, 'as of the sound of many waters'.

In contrast to these urgent eruptions around him, Thoreau's was a quieter resistance – for all that he practised civil disobedience and even went to prison for one night as a protest against his taxes being used to support war and slavery. If he was said to be an ugly man, he became beautiful by virtue of what he observed and absorbed. 'I went to the woods because I wished to live deliberately, to front only the essential facts of life …' He too discovered slowness, seeking meaningful leisure 'for true integrity', free of property and employment. At Walden, he found the earth's eye. And as he looked into the water, he saw the rest of us too. His last words, knowing he was about to die, were 'Now comes good sailing.'

Exiled on the island of Crete, Daedalus was imprisoned by the waves and determined to escape. The only way was up. The architect made wings from feathers fastened with wax, hindered in his work by his playful young son. As they set off on their timeless flight, Daedalus told Icarus to ignore Orion the Hunter and be guided only by his father. He must not get too close to the sun or the sea; either would be disastrous.

They took off into the air, watched by fishermen and shepherds who thought they must be seeing flying gods. But Icarus could not resist going higher; perhaps he wanted to be a god himself. Rising high in his ecstasy, he left his father far behind and soared up towards the sun. As his wings melted, his dreams came apart and the boy tumbled into the sea. His father called for his son in vain. Then he saw the feathers on the waves, and cursed his own invention.

Something falls out of the sky and into a lake. There's a burst of white against the blue. His coming is seemingly unnoticed by the rest of the world. He wanders into a one-horse town; the landscape is as arid as the place from which he came. He looks like a comet, his flame-like hair slicked back on entry into the earth's atmosphere. Even stumbling down what appears to be an enormous cinder pile he is otherworldly, a pale bird of paradise picking his way through blackened coals. As the visitor exchanges his grey overalls for a plain black suit, he assumes a worldly guise and an ordinary name and proceeds to build a secretive corporation, World Enterprises, producing inventions from instant cameras to electronically generated music.

A lavish art book lands on the desk of a university professor, Nathan Bryce. It is published by WE and entitled *Masterpieces in Paint and Poetry*. It falls open – to a burst of seagulls on the soundtrack – revealing Brueghel's *Landscape with the Fall of Icarus*.

A mere splash of a boy, all kinked knees, plunges into the corner of a world going about its business. A ploughman's red shirt draws the eye to the centre of the picture; a ship sails on a Renaissance sea. On the facing page are Auden's lines:

> ... the sun shone
> As it had to on the white legs disappearing into the green
> Water, and the expensive delicate ship that must have seen
> Something amazing, a boy falling out of the sky,
> Had somewhere to get to and sailed calmly on.

All these scenes seem to refer to something else before and after, as if, like *Orlando*, the story was running backwards and forwards in time. As Mary-Lou the hotel maid shows Thomas Jerome Newton to his room, he collapses in the lift. She has to pick him up in her arms – as thin as he is – carrying him down the corridor, his nose bleeding, head lolling and legs dangling like a pietà. The next day she and Newton go to church, where they sing 'Jerusalem' – *And did those feet, in ancient time, walk upon England's mountains green*.

Newton is oddly innocent, both in and out of control, invincible and vulnerable, wide-eyed at the world yet knowing it all. He is a Prospero arriving on island Earth; a refugee from Anthea, his planet from which the sea has disappeared. He is also an omen, a saviour and a sacrifice: seeking to save his own home and warning us of the danger to ours. He might as well walk on water; even his centre-parted hair is reminiscent of Jesus. In the novel by Walter Tevis on which the film is based, the comparison is more obvious. When he sees a painting of Christ on the cross, Newton is startled – 'with its thinness and large piercing eyes it could have been the face of an Anthean' – and when Bryce wonders if there will be a second coming, the alien replies, 'I imagine he'll remember what happened to him the last time.'

The way he inhabits the film is unsettling. His performance has an interior, kabuki air – perhaps because he had been trained in mime by a man who had felt 'confronted by this vision of the beautiful archangel Gabriel, glowing, shining' (among the star's improvisations were sailors drowning at sea and animals hunting their prey; he and his tutor-lover, Lindsay Kemp, also planned to make a musical of The Water-Babies together, just as two years later, Derek Jarman wanted him to sing Ariel's songs in his film of The Tempest). On location he kept to his trailer, behind its wooden blinds, declining to mix with the rest of the cast; he told his director, Nicolas Roeg, that he wanted to play Newton as a recluse like Howard Hughes, although he also became addicted to eating ice cream, a habit that had to be curtailed since he started to put on weight, which may explain the corset he wears in later scenes.

Tevis describes his character's bones as bird-like; when he is driven about, Newton tells his chauffeur, Arthur, to slow down and keep to thirty because he is unused to earth's gravity. And in an operation in which he reveals what appears to be his true self, he pulls off prosthetic nipples and penis and takes out his lenses to reveal yellow eyes with vertical slits, like some creature looking out over the veldt. It is another mask. All the time, we are aware that the real person behind those blue eyes – which were in reality odd enough, one permanently dilated pupil ringed with bird-like intensity after a boyhood fight – was a transformer who had moved through animal as well as human guises. Most lately he had become half-canine, half-human, a dog star, dangling a diamond earring and singing with a feral yelp, prowling a post-apocalyptic city whose skeletal wreckage prophesied the future ruins of other towers. Like the wolfish Virginia's vision of Vita as a porpoise or a stag, he was fabulous in the true sense of the word, since fables are our fates as embodied by animals.

To me, he became another creature too. Looking out over the lake where he fell to earth, Newton sees his desert planet and his family walking through endless dunes where the sea had been but was no more. He dreams of aerial oceans, and in a sequence which I only registered subliminally then, he and his Anthean wife spin around one another in a fluid cybersexual space while we hear snatches of whale song and 'oceanic sounds' supplied, as the credits now reveal, by Woods Hole Oceanographic Centre, Cape Cod, and recorded by Frank Watlington.

During the nineteen-fifties, Watlington had been working for a secret American base on Bermuda, where he devised an underwater microphone to listen for Russian submarines, in much the same way that astronomers would listen for signs of extraterrestrial life. Having lowered his hydrophone fifteen hundred feet into the Atlantic, he recorded the songs of humpback whales for the first time. Watlington had kept his discovery secret for fear it would be used by whale-hunters. But in 1967 he drew it to the attention of a biologist, Roger Payne, and his wife Kathy, who together with the scientist Scott McVay realised there was a repeating pattern to the strange noises off the island that had inspired *The Tempest*. The Paynes began to record the songs for themselves from their yacht, *Twilight*. 'Far from land, with a faint breeze and a full moon, we heard these lovely sounds pouring out of the sea.'

In 1970, Roger Payne released *Songs of the Humpback Whale*. The future had been announced – not from beyond the earth, but beneath its seas. The blue-toned album cover, showing a whale breaching out of the ocean, echoed another soundtrack – 2001: *A Space Odyssey*, released two years earlier, displaying a spacecraft shooting from its mother ship like a slender cetacean. Payne's hydrophone recorded an animal out of time and space. We were looking for aliens beyond our galaxy, when all the while they were living in

our oceans. Even their binomial, *Megaptera novaeangliae* – big-winged New Englanders – invoked a new age. We'd seen our blue planet from outer space and realised that there was nothing we could do. When Newton tells Bryce, 'Our word for your planet means "planet of water",' he is citing Arthur C. Clarke, the author of 2001, who thought the Earth would be better named the Ocean.

In 1972, an atomic-orange poster for a concert to save the whales showed the starman as a space-age Ariel astride a grenade harpoon, superseding its barbaric cruelty with his lamé and lurex. Around the same time, I saw my first whale: a stolen animal in an inland pool, the same graphic orca that would leap out of my note-book. It too was a sign, although I didn't understand it. It wasn't until 2001, shortly before the aeroplanes dropped out of the sky, that I saw another whale, launching itself off Cape Cod like the rockets which had blasted off from Cape Canaveral in the seventies, sending probes spinning into the solar system – lonely voyagers loaded with whale song on golden discs for aliens to hear, just as John Dee used his golden disc to communicate with angels, and just as Kubrick's film hums with another alien sound, emitted from a mysterious black monolith. Like *The Tempest*, 2001 is filled with classical references and performed in formal language as a masque, a ritual moving languidly from the dawn of man, via titanium-white spaceships, to the end of time.

It is also, as Andrew Delbanco observes, a very Melvillean film. The astronaut, David Bowman – his name oddly close to that of the performer who would be inspired by the film to write a song that set him adrift – resembles a beautiful version of an American football player as he lies under a sunlamp in his white shorts. Space pods float like bathyscaphes; they might as well be sinking in the benthic ocean, crewed by deep-sea divers. Later, Bowman's mate, the orange-suited Frank Poole, is lost in fathomless space, spinning into infinity.

Bowman, strangely distanced from the drama into which he is plunged, is a twenty-first-century Ishmael caught up in an unknown mission; not for a white whale, but for a black slab.

Indeed, a generation before, just before I was born, John Huston had come to Britain to film *Moby-Dick*, his nineteen-fifties version of Melville's science-fiction search for a grand hooded phantom swimming in a black ocean, spanning time and space, tugging Ahab out of his boat and into the infinite sea.

All these things seem so fated, from this distance and this close, that I'm not surprised to find that all three alien creations – black monolith, fallen angel, white whale – were filmed in the same studios outside London, born in a sort of suburbia, out of imaginations contained behind tree-lined avenues.

I first saw Kubrick's film in a small cinema in Southampton's nineteen-seventies high street, barely recovered from its wartime blitz. The narrow building, more accustomed to showing Swedish porn, was slotted between a shoe shop and a bank; a secret place down a dark lobby, more like a nightclub.

I entered as if I knew I were doing wrong.

Two hours later I emerged dazed from the hallucinogenic ending – the astronaut's blue eyes staring and shaking as they witnessed the making and unmaking of worlds, blinking in a white room like a giant microscopic slide where he lay as an old man, wondering if that was the way to die, only to become a star-child in a caul, floating far above the earth.

What remained with me, and still does, is the deepness of these worlds – the spaceship, the alien, the whale – only waiting to confirm our future. Whiteness was its aesthetic, an appalling pallor, a plastic dystopia fabricated from petrochemical products that were already starting to poison us. Now, fifteen years after the passing of the year it commemorated, the slowness of Kubrick's film and its sparse words seem more Jacobean drama than modern movie; a Cinerama version of Prospero's story, with its Caliban apes (led by Dan Richter, a Provincetowner), its computerised Ariel, Hal, and its brave new world. These moving images barely move at all; continually showing in a loop in some decrepit cinema since I first saw them, they are mechanical, needing and creating light, dreams burned into and out of their celluloid.

How small the past looks from the future; even yesterday seems old-fashioned compared to tomorrow, and the day after that. In my Provincetown attic, as the sun sets over the bay, I re-view Roeg's film, which I first saw forty years ago, watching it on my white computer, which might have been made by World Enterprises. Newton, wearing a suit of long underwear and a grey buckled corset to keep his bird bones intact, sits in front of a bank of televisions. This is the way he has learned about the earth back on his planet, picking up the incontinent transmissions of the human race as they leak out into space, sampling images to be watched by aliens as if through compound eyes. It was the same way I'd learned about the world, too, from the TV in the corner of our front room. Then, machines were the future. Now, the space station slides across our suburban skies, night after night.

Each flickering cathode-ray tube, tuning in and out of Newton's array, broadcasts its own footage: an Elvis movie, lions mating, a NASA rocket launch. But in a cut-up worthy of Burroughs or Paolozzi, there's one black-and-white scene that causes me to pause the film, something I could once have done only in my head. It's as though the director-magician has inserted it retrospectively for my attention. With the dark sea roaring outside, it seems to have been conjured up out of my wishful thinking.

For a moment I wonder if I've imagined the whole thing. I rerun it, rerunning time, and watch it again.

In snatched clips we see a young Terence Stamp, an Adonis with bleached blond hair playing Billy Budd, the Handsome Sailor, in a sequence taken from the 1962 film of Melville's novella. (The director cannot say why, he later tells me in a London basement.) Betrayed by Claggart, the punitive master-at-arms whom he has accidentally killed, the angelic Billy is about to die. Captain Vere declares solemnly, 'If you have nothing to say, the sentence of the court will be carried out' – and we see the sailor with the rope around his neck crying, 'God bless Captain Vere.' As Billy's body is about to be consigned to the deep, the crazy images collide and Newton responds in panic as if he'd seen his own future, 'Get out of my mind, all of you!'

That summer of 1976, I left school. The days extended, ever hotter, falling on their knees. Southern England became a landscape devoid of water. The port city where I lived registered the highest temperature in the country. The grass burned and the earth baked, great cracks opening up as if to swallow me. If I'd thought about my future at all, I would have felt sorry for the boy I was rather than the person I would become. I wore my sunglasses and my jumble-sale clothes as a collage of all the people I wanted to be, gathered from the dead of a generation whose belongings were being disposed of in scout halls. I felt the texture of the past, recycling history that had never happened to me, looking forward to a foreshortened future. I lived in another time, another place. I don't suppose it occurred to me that the national emergency – when the rain failed to fall, reservoirs dried up, fires raged over heaths and crops withered – was a sign of things to come. We waited for some change in the weather. Newton's dry planet was my own. Water had acquired a new meaning, but I had no idea. I couldn't swim.

At the beginning of the year I'd stood on a railway platform dressed in white school shirt, salvaged black evening waistcoat, trousers with satin seams and cracked patent shoes, my hair slicked back and sprayed gold, waiting to be conducted to a pitch-black arena where Kraftwerk's crackling 'Radioactivity' and the brutality of *Un Chien Andalou*, in which an eye is sliced open with a razor, gave way to the sound of a relentless engine and the sight of the man himself, improbably teleported to an earthly stage, sunken-cheeked, too thin to be contained by the fluorescent cage of striplights thrown around him.

He'd arrived from Dover in a slam-door train, to be driven from the station in an open-top car like someone to rule us. In a bankrupt, blacked-out country with the power cut off and insurrection threatening, the whiteness of 2001 seemed far away. I felt I had been summoned to his presence, to the future. He was in a new incarnation, a self-portrait seen in a dark glass. There he was, dressed in black and white, a pack of Gitanes in his waistcoat pocket, a Prospero throwing darts in lovers' eyes as he quoted from *The Tempest*, singing of the stuff from which dreams are woven; bending sound, dredging the ocean, lost in his circle. I was alone with him in the darkness, overlooking that ocean. The cavernous auditorium in which he performed his apocalyptic cabaret was built as a swimming pool for the 1934 British Empire Games, two years before other games were conducted in another European capital – cities connected by that same railway, that same past and future, that same deathly glamour. I had no idea that his opening song was inspired not by trains but by the Stations of the Cross; although I always noticed the two crucifixes around his neck.

We were separated by the sea: the Pacific he overlooked from his room and the Atlantic he crossed in white liners; New York and California, Hockney's pools and Max's Kansas City; white boys in peg-top trousers and dyed hair and chic black girls with peroxide crops. Stranded on a south-coast estuary, mine was a world of wood-trimmed trains and slow-moving cars and suburban streets which still seemed to operate under the shadow of rationing. He gave me a dark glittering world, his monochrome figure held against the moonlight, mime-walking as if caught in a loop, marking time. We'd come from the same place; I was physically like him; yet in that flashlit year of transition he had been transformed anew, by the same alchemical process which had burned his hair.

Gods require us to believe in them; they wouldn't exist otherwise. Seeing him in the flesh was as disconcerting as the killer whale in a suburban pool. I needed to close that distance. I wanted to feel the high collar of his white shirt against my neck; the tailoring of his waistcoat around my ribs; to look normal, like him, and yet utterly different, like him. I came out that evening carrying the tour programme, ISOLAR, a cross between a newspaper and a secret manual, its cover with his back turned on us like a priest. I looked for every coded detail in its pin-ups set alongside inexplicable radioactive images of his forefinger and a crucifix and a photograph of him in dark clothes daubed in diagonal stripes of white paint from his slash-necked top to his rolled-up trousers (taken in at the waist to fit his shrunken frame) down to his socks. He stands straight, arms by his side, looking to the horizon, disturbing it like a dazzle ship, moving in and out of register like a mistuned television. Even in yellowing newsprint he's still performing, stripped down to a DIY aesthetic, as if he was a teenager like me, dressing up in front of his bedroom mirror.

And as my first year of adulthood began with witnessing him on stage, it ended with seeing him on film, more remote as he became more intimate. In the sex scenes, his hairless body is both elegant and awkward; he appears hermaphroditic, both animal and disembodied, as if his lower half had always belonged to a dog or a deer. His pale face was his fate, and mine.

Now living by the lake where he first appeared, Newton comes to Bryce one evening, appearing at the end of a jetty: an ashen, hooded doppelganger in a duffel coat, looming out of the darkness like a ghostly monk.

'Don't be suspicious, Dr Bryce,' he says, before disappearing back into the night.

In the chapter in Tevis's book which inspired this scene, Bryce walks around the lake towards his employer's house, a white nineteenth-century clapboard mansion, and stops to stare at the water. 'He felt momentarily like Henry Thoreau, and smiled at himself for the feeling. *Most men lead lives of quiet desperation.*' Bryce sees Newton walking towards him, wearing a white short-sleeved shirt and grey slacks. As he draws steadily nearer, the alien's otherness is made manifest. 'There was an indefinable strangeness about his way of walking, a quality that reminded Bryce of the first homosexual he had ever seen, back when he had been too young to know what a homosexual was. Newton did not walk like that; but then he walked like no one else; light and heavy at the same time.' He seems to move as if through water.

The two men sit and drink wine. Bryce wonders whether Newton is from Mars or Massachusetts, and he thinks of the migration of birds, 'following old, old pathways to ancient homes and new deaths'. He's worried that the spaceship he is helping Newton construct may be a weapon. Newton asks Bryce if nuclear war, which caused his own planet's devastation, is imminent (Tevis's book was published in 1963, a year after the Cuban missile crisis). All the while, as the film makes clear, dark forces are marshalling against the stranger in a country once settled by visitors but which now sees them as a threat. 'This is modern America,' they murmur, 'and we're going to keep it that way.'

Bryce returns to his house drunk, and peers at a reproduction of *The Fall of Icarus* on his kitchen wall (Sylvia Plath had the same picture hanging in her apartment). He sees 'the sky-fallen boy' who 'burned and drowned', and wonders if Daedalus' invention was where it all went wrong. In the film, Newton stands by a blue light on his jetty overlooking the lake, his incandescent locks and pale profile held against the water, later to reappear on another nuclear-orange cover. He's an Ishmael or a Gatsby, sensitised to psychic disturbances in the way that Scott Fitzgerald's mysteriously wealthy anti-hero is 'related to one of those intricate machines that register earthquakes ten thousand miles away'.

And in a sequence which plays and replays over and over in my head, Newton is being driven through the backwoods when he sees a group of early settlers in sepia, as if his vision had become doglike, as if he were seeing through a filmy continuum. He looks out; they look back, amazed to see, not a boy falling from the sky, but a car driving past. And we see it all in the way we see scenes from a train, as Virginia Woolf wrote, 'as a traveller, even though he is half asleep, knows, looking out of the train window, that he must look now, for he will never see that town, or that mule-cart, or that woman at work in the fields, again'.

'I'm not a scientist,' Newton tells Bryce. 'But I know all things begin and end in eternity.' It was another reference, as Roeg would reveal, to William Blake, who saw angelic figures walking among the haymakers in Peckham Rye and trees filled with angels as if caught in the branches, and wrote, 'Eternity exists, and all things in eternity Independent of Creation which was an act of Mercy.' And I think of Blake's image of another Newton: the alchemist-scientist sitting at the bottom of the sea, futilely measuring out infinity. Centuries before – although it might have been hundreds of years ahead – another magician, Prospero, asked Miranda, 'What seest thou else | In the dark backward and abysm of time?' That year, recording in a French château, the star asked his producer what a particular piece of studio hardware was for. He was told, 'It fucks with the fabric of time.'

He scared and excited me. I wanted to be him, not to have him. He represented freedom and danger. He was absolute artifice, but utterly feral too: nothing could be so exciting or so challenging. He had only to walk into an Amsterdam hotel to appear completely otherworldly; wherever he appeared, he was the focus of disruption, a disturbance in the ether. He was my dark star. He had summoned me to the city. A year which began for me watching the television in the suburbs – the same screen through which I first saw him, posturing as a corrupt, tinselled Nijinsky, pawing at his guitarist like some predatory animal – ended in my descent into a real nightclub. I slammed the train door and crossed the river, a brown god by day, a black serpent by night, and walked down a dark street surrounded by empty-eyed warehouses to stand in a queue where a boy in a biker jacket with spiky bleached hair asked me for a light.

As the flame sparked in his face he became another Ariel. We were children of a man stealing time who told us that nothing could help us and who, having visited an oceanarium, as his friend dimly recalls, wished he could swim like a dolphin, like the dolphin he would have tattooed on his skin.

When *Songs of the Humpback Whale* was released, some people believed they were listening to the voices of aliens. In Walter Tevis's novel, the description of the recording of Newton reading his poetry in Anthean might as well be an account of whale song, 'sad, liquid, long-vowelled, rising and falling strangely in pitch, completely unintelligible'. In the book, Newton – who, it is now clear, came to rescue humanity from nuclear and environmental destruction – is blinded by the inept scientific examinations of the authorities, 'certainly not the first means of possible salvation to get the official treatment'. The alien admits to being afraid of 'this monstrous, beautiful, terrifying planet with all its strange creatures and its abundant water, and all of its human people'. In the film, Newton is released by the mysterious agency from his apartment-prison, with its rococo wallpaper depicting those same uncanny beasts. In the last scene, set in some future Christmas, we see him – his herringbone coat over his shoulders, his burnt-out eyes shielded by sunglasses, his flaming hair held under a fedora like the ghost of Christmas past and yet to come – drinking in an outdoor bar.

There he is tracked down by Bryce, now an old man. Only the alien remains the same; the rest have aged around him, although he has become human, like us. He has released, anonymously, a grey-sleeved album, *The Visitor* (when Bryce buys it in the supermarket, we see the colourful cover of *Young Americans* displayed behind him). As they talk, Newton toys with his drink – his thirst having become an addiction, binding him to earth – and he tells Bryce that his recording has a message for his wife which he hopes she will hear when it is played on the radio, broadcast like whale song, to be picked up back on his home planet. But as he looks up from the table to a helicopter whirring overhead, it is clear that the grounded angel will never be set free; that he will always be under surveillance, like the rest of us.

When interviewed about this time, the star said his work was about the sadness he felt in the world. Perhaps that was why, in some performances on his tour, he appeared with a thin gold veil over his face, denying the gaze of thousands of eyes. A veil to stifle as well as conceal. It was a strangely silent age, for all its noise. Sound had to be physically summoned through a needle or read from magnetic tape.

There was something about this breakpoint, three-quarters of the way through the century, the year in which I turned eighteen. There seemed to be a disturbance in history, the result of accelerating time; the threat of a disrupted country and an overturned world, the threat of violence and extinction, the last point at which the planet was still sustainable; the shift from the twentieth century towards the twenty-first that exposed things between the cracks.

When filming in New Mexico – like New England, a place cleansed of its first peoples – the production used locations which included a Mesoamerican graveyard, from which the actor – who was now living on milk, red peppers and cocaine – insisted on having his trailer moved. Meanwhile the crew reported malfunctioning cameras, an echo of Bryce's attempt to covertly film his employer using a World Enterprises x-ray camera, only to find that the resulting image showed an empty body.

Throughout this period of flux the star felt haunted by the personae he had created. Beset by a metaphysical terror, himself an exile in America, he consorted with the supernatural, drawing pentacles in chalk and burning black candles, discerning the shape of the devil in his swimming pool and believing that bodies were falling past his window in Los Angeles – a city of lost angels – just as in the film's most horrifying sequence, Newton's gay lawyer, Oliver Farnsworth, is thrown through his own apartment window by secret agents. The men pick up Farnsworth by his arms and legs, swinging him at the plate glass, but his body bounces back and he apologises to his assassins, one of whom replies, 'Oh, don't worry about it.'

Now at college in an eighteenth-century gothic building, exiled in a London suburb – neither part of the city nor my home – I went to see the film over and over again, as if I were going to Mass rather than the movies. I did not know then that it had been made in studios only a few miles down the road. I walked to the local cinema along the river that followed me down Cross Deep, ignoring its wide green waters that ran under drooping willows. I had no interest in it, only in the nineteen-twenties building ahead, white-tiled like a temple. I took a cassette recorder from my shoulderbag and, stashing it between my feet, illicitly taped the soundtrack so that my underage sister could hear it. The tape did not come out blank, but I might as well have been recording ghosts.

That time is thick with memory. Back in my cell-like room, while my peers played on the football pitches outside, I stuck plastic cups stolen from the refectory in a grid on the wall above my bed to emulate his soundproofed capsule. I fetishised what he wore and tried to wear the same: the high-waisted quilted trousers that ended below his knees in scarlet turn-ups; the plastic sandals and seventies shirts patterned like his wallpaper; even the green visor, white shirt and belted shorts in which he plays table tennis in a room surreally decorated like a forest with autumn leaves scattered on fake-grass carpet. I dyed my hair red, using the same dye my mother used, secret silver sachets from the chemist, and wore a green woman's jacket with padded shoulders from the nineteen-forties.

To me he represented all the weirdness of that decade which he created and which created him. I didn't know what people were, physically; I thought then, and still think, that they were another species altogether; that it was they who were the aliens, not me. In the illusory intimacy of film, I thought I was seeing him as he was in everyday life, although what had been created was something stranger, a stranger in a strange land. His melancholy was mine. He told me so.

This is how I remember all of this; it is not necessarily how, or why, it happened, forty years ago, or forty years hence.

'He's just a man,' friends tell me. He may be now, and he may have been before, but he wasn't then.

When he was photographed for twelve hours in a Los Angeles studio, continually changing into different outfits, he might as well have turned up on Napoleon Sarony's doorstep in Union Square with a suitcase full of costumes, as Wilde had done when he came to New York as an alien. A century later the starman would arrive on that same square, now home to Warhol's Factory, wearing a floppy hat and long hair like an Apostle of the Newness, auditioning for another role. Or perhaps just reprising it, in the array of identities which led to that point (and then to another, and another). A few months before he had appeared on a record sleeve in a silvered 'man dress', reclining in a pose that referenced Wilde's languid form draped on Sarony's couch.

Night after night, as I came out of the cinema and into the sodium-lit street, I thought that the star – who spoke of his isolation and near madness – would walk past me, with his wide urgent stride, wearing his plastic sandals, as if he'd fallen to earth in the nearby river.

'Surely all this is not without meaning,' says Ishmael, as he stands looking out over the water while Manhattan rises and falls behind him. 'And still deeper the meaning of that story of Narcissus, who because he could not grasp the tormenting, mild image he saw in the fountain, plunged into it and was drowned. But that same image we ourselves see in all rivers and oceans. It is the image of the ungraspable phantom of life; and this is the key to it all.' 'She had been a gloomy boy,' Woolf wrote of her Orlando, 'in love with death, as boys are.' We all shudder at our earlier selves, and sigh with regret at what we did or did not put them through.

Not long ago, a museum curator led me down into the belly of the building, to the conservation studio housed in its basement. Staff were busy tending to various dummies dressed in glamorous outfits from the twentieth century, each being prepared to take its place in the galleries above. In one corner of the room a plywood capsule lay propped on trestles, like a bier. The curator lifted its curved lid and inside, under a cloud of tissue, lay a costume shaped from tarnished silver gauze; a frail, glittering suit of armour studded with sunflowers and leaves at the shoulders and hips for some Wildean parable. It had been worn in a film in which the starman played another alien, a pierrot picking his way over an irradiated beach lapped by a solarised sea. His long white stockings lay beside the outfit, rubbed with dirt.

Boxed up and ready to be launched out into the ocean, stiff on its legend and sewn over with sequins and pearls, the outfit all but trembled as I looked at it, as if an aquatic insect had emerged from its chrysalis and flown away. It might have been made of nothing. I could have reached down and touched it, this phosphorescent shroud shaped with his form, no more substantial than the glowing mantle on a gas lamp. But I couldn't make that final connection, for fear my dreams would disappear. It was too late. The moment had passed, forty years before.

Recently I heard his voice, inadvertently, on the radio in my kitchen. It was an interview from 1976 and he was speaking as a civilian, as an ordinary person, making jokes. As I walked past, it sounded so strange and familiar, so confiding that for a moment I thought I was listening to myself.

On 10 December 1972 the starman, who declined to fly and travelled instead by sea, left New York on RMS *Ellinis*. On his way across the Atlantic, sitting in an overstuffed armchair in his suite, he read *Vile Bodies*, which opens with a stormy passage on a ship. Inspired by the book's sense of futility and its passionate bright young things caught between the wars, he wrote a song subtitled

'1913–1938–197–?' I remember not wanting to acknowledge the date. On 21 December he sailed up Southampton Water, from where I watch other liners pass like glamorous ghosts in the night. I ought to have been there to greet him when he arrived, to see his peacock plume coming down the gangway, although I was only fourteen and still at school. But forty years before, on that same quayside, I might have met one of those figures from Waugh's novel, making the same journey in reverse.

The Honourable Stephen Tennant was born far from suburbia, into a life of privilege and landed estates with intimate connections to power and influence. A magazine cutting torn from his journal shows his family on an earlier visit to America, in 1919. They had sailed from Southampton to New York, and travelled by private Pullman train to Boston, where Stephen's mother consulted the city's mediums, trying to contact her eldest son, who had been killed three years before on the Western Front.

TINGUISHED VISITORS FROM OVERSEAS AT THE
SANTA BARBARA

rd and Lady Glenconner of Lon
David and and Stephen
finished

the Right Honorabl
first bungalow

Stephen's father, Lord Glenconner, wearing an extravagant bow tie, sits secure in industrial fortunes gathered in the previous century, and in the privilege of his position. In 1914 it had been Glenconner to whom the Foreign Secretary Sir Edward Grey turned and said, 'The lamps are going out all over Europe; we shall not see them lit again in our lifetime.' Next to Glenconner, in a picture hat and tea gown, is his wife Pamela, product of a more poetic breed, the Wyndhams; as a young woman she had known Wilde. David, Stephen's brother, wears a dark suit, his raffish future as a playboy and nightclub owner presaged by his suede shoes.

But it is Stephen who draws the eye, for all that he too was only in his fourteenth year. It was impossible for any camera to take a casual picture of him. He predicts himself, demanding attention through his own discerning eyes. He stands under a palm tree in his white shirt and wide tie, his trousers cinched with the kind of snake-clasp belt I also wore as a boy. On the family's tour of America he gave dancing demonstrations in hotel ballrooms and painted pictures in a Beardsley style; his mother believed her son was spiritualistically inspired by Blake. And from the beach he collected abalone shells, 'the kind with bluegreen iridescence'. They would become an obsession for Stephen, as if their nacre held the sea just for him.

By his coming of age in 1927, a very different one from mine, Stephen had reached the peak of his perfection. That year he was photographed in his Silver Room in the family's townhouse in Westminster, where a streetlight illuminated the foil-covered walls and ceiling within, and dull silver satin curtains fell like waterfalls. A polar bear skin was splayed across the floor, parrots perched in a silver cage, and alligators languished in a glass tank. This icy reflecting chamber resembled an aquarium installed in a Georgian square. It was a reservoir of Stephen's dreams. He might as well have kept a porpoise there too, fished from the frozen Thames.

Against a silver backdrop, the colour of water lit by an electric moon, Stephen posed for Cecil's camera. Clad in a pinstripe suit, striped shirt and silk tie, he wore a shiny mackintosh, enhancing his mercurial sheen. It was the most outrageous act of cross-dressing.

Stephen had assumed the conventional attire of a City gentleman. Never had pinstripes seemed so unconfining, no tie so decadently tied. Their ordinariness was stranger than his made-up face and halo of lightened hair. He was an alien in Mayfair, a star of his own making, posing for an album cover in a room as fabricated as Warhol's Factory. In public he appeared in pageants as a pearl-clad Romeo or as Shelley, 'gold-dusty with tumbling amidst the stars'. But in more intimate entertainments, Stephen and his friends were filmed all the while by a tall young footman in dark glasses, rehearsing scenes yet to come.

In retrospect 1927 seems a signal date, caught between what had happened and what was to happen, as if history could go either way. That same year Virginia Woolf created her Orlando, whose conflation of glamour and gender mirrored Stephen's: ageless, aristocratic, fluid and vulnerable. The endless rooms of

Orlando's palace are awash with exquisite things from the past, 'when the silver shone and lacquer glowed and wood kindled; when the carved chairs held their arms out and dolphins swam upon the walls with mermaids on their backs'; and to animate his domain, 'he had imported wild fowl with gay plumage; and two Malay bears, the surliness of whose manners concealed, he was certain, trusty hearts'. (Vita was once given a Russian bear cub; it had to be put down.) One room is entirely silver, down to the counterpane, although Orlando considers this a little vulgar.

Nor was it by chance that in 1927 Rex Whistler portrayed Stephen as Prince Etienne in his Tate Gallery mural *The Expedition in Pursuit of Rare Meats*, 'an Arcadian strip cartoon', as his brother Laurence called it, with its anachronistic figures in eighteenth-century dress riding bicycles, their hair Eton-cropped. Stephen's legend was being played out ahead of him. To further honour this dauphin's adulthood, which would never really arrive, Stephen was sculpted by Jacob Epstein, who was not sure if he was a boy or a girl, but portrayed him half-naked, much as the starman had been photographed as a diamond dog. As Stephen watched himself taking shape in the artist's hands, he declared, 'this exquisite grey creature looking like a drugged, drowned Parsifal ... when I am dead & forgotten its loveliness will live, gazing back into the past at me – where Ghost meets Ghost'.

Although they seemed not to have caught what Cocteau called 'the illness of time', all too soon the passionate bright young things burned themselves out. Stephen and his lover Siegfried Sassoon escaped to Sicily, a legendary island, where Stephen lay about in satin pyjamas while Siegfried, who had once fought the Germans in the mud of the Western Front, cleaned the shells they'd gathered from the beach. Stephen felt that they contained their love. But they became empty echoes when he had to send his lover away; Stephen could never bear too much reality. By 1935 he had retreated to his Wiltshire home, Wilsford Manor, where the river ran clear through his garden and the water meadows beyond. Restless, lovelorn, he wandered to Sussex, to visit Virginia Woolf at her riverside house

and its flooded chalk valley. Woolf saw him often that summer and found his attentions amusing, although she looked upon him as a 'fish in a tank', so Stephen Spender told me.

What did she see in Stephen? Her letters and diaries of the time follow him and his like with a kind of amused disdain, filled with scenes of the half-life of high society and its shimmer. In one of her albums, along with all the photographs of her and Vita and their spaniels, and Lytton Strachey and Tom Eliot on English summer days, there is a caption in her hand – 'Stephen Tennant' – but like those nude photographs of her and Rupert Brooke, the picture is missing, edited out by history, and there is only a blank space where Stephen used to be.

Yet they were not so far apart, these two, attuned to their inner loneliness, fragile in the face of the modern age. Virginia was forced to stay in bed in order to control her mental illness; Stephen, having suffered from consumption as a young man, chose to stay there: for both, such instability seemed a reaction to the speed of the twentieth century. When she returned to Rodmell after lending her house to Stephen, Virginia wondered if she'd find the bathroom littered with his cosmetics; but he had moved down to the coast and the Bay Hotel, set on the cliffs at Seaford. He bought a new journal in the town, and on its cover painted a neo-romantic seascape overlooked by a classical bust and a sailor hailing from a placid quay.

The reality was somewhat different. Seaford was known for the huge waves that crashed over the sea wall in autumn and winter, smashing windows and damaging buildings. In the early hours of 17 September 1935 a terrific storm, the worst for many years, swept down the Channel. In the tempest, all fancies were forgotten. How could anyone sleep or even live through such a storm?

Stephen was woken by the sound of shattering glass. Rushing out to the landing, his plush monkey in his hand, he looked down on the moonlit scene as the sea lashed the shore. 'I sat in the upper landing window – & watched vast ghosts form & dissolve on the esplanade – waves, giant waves.' The winds, roaring at one hundred miles an hour from Folkestone to Plymouth, tore boats from their moorings even in protected Southampton Water, and claimed several lives. The next day Virginia drove down to Seaford to see the waves exploding over the lighthouse; she was prevented by the police from driving along the flooded road. Meanwhile, along the coast at Brighton, the doors of the Victorian aquarium – a subterranean chamber where porpoises, a beluga whale and two manatees were kept in the nineteenth century and where, in the nineteen-seventies, I would watch dolphins perform in a space resembling an underground car park – were burst open by tons of seawater and shingle driven in from the beach, flooding its interior six feet deep and all but allowing its inmates – among them a twenty-seven-foot-long stuffed basking shark – to return to their rightful homes.

Throughout the night Stephen's 'transparent dreams' were assailed; the windows banged and the rain lashed the ground, and his calm was only restored in the morning, when the bright light of day brought a letter from Morgan Forster, who was coming on a visit. A week later, having moved inland to Forest Row to escape the raging sea, Stephen sent his Rolls-Royce to the station to meet his friend, who may or may not have been surprised to find bouquets of roses, gladioli and carnations arranged inside, with a pair of tropical butterflies perched quiveringly on the flowers, their frail beauty Stephen's gesture to the aftermath of the storm.

As an 'aquaeous' darkness fell outside the dining room, the two men talked late into the night. They spoke of Stephen's evening with T.E. Lawrence in his hut-like cottage in Dorset; Morgan liked Stephen's description of the hero of Arabia talking in 'morse code'; the memory was more acute because Lawrence had since died after an accident on his motorbike. Morgan recounted his recent dinner with Noël Coward, whose equally staccato voice he imitated, and recalled a weekend in Amsterdam with Christopher Isherwood, Klaus Mann and Brian Howard. Then the two men fell to discussing *Moby-Dick* and the character of Queequeg, the tattooed savage, who fascinated them both. In an essay published in 1927, Forster had called *Moby-Dick* a 'prophetic song' beyond words, and saw *Billy Budd* as a 'remote and unearthly episode', its sadness indistinguishable from glory. For his part, Stephen's love of Melville merged into the photographs of sailors he stuck in his journal and the fantastical Lascars of Marseilles he drew alongside them, characters from Jean Genet out of Jean Cocteau.

The convivial evening ended with Morgan playing the piano and reading aloud from his stories. But when he announced he had to go, Stephen incurred his displeasure by quoting Virginia, who had complained, 'Morgan's always in such a hurry to leave me.' Stephen could not imagine why anyone would want to leave him.

On 20 November 1935, Stephen sailed from Southampton on the White Star liner *Aquitania*. He was bound for America, a journey Virginia had been able only to imagine. A century before, Robert Browning had written, 'there is but one step to take from Southampton pier to New York quay, for travellers Westward'. The transition was seamless for Stephen, whose home travelled with him, carried into that great white vessel like the starman's wardrobe.

He took: a new blue jumper with red zips by Schiaparelli, a powder-blue hussar uniform with silver braid and skin-tight breeches, and a sensible herringbone suit for the country (Rex Whistler told him it looked like the kind worn by men who said, 'Well, I'm going to turn in'); a travelling library that ran from a biography of William Beckford to Sylvia of Hollywood's beauty tips, *No More Alibis*; numerous cosmetics, perfumes and journals; and a little 'Jane Austen' cabinet containing his most precious shells. Stephen had only reluctantly turned down a suggestion by Peter Watson that he should take his new Rolls-Royce as well.

Six days later, after an unsteady voyage installed in a stateroom on A Deck – Rex was travelling on B Deck below – Stephen arrived in Manhattan and was swept from the honking, bustling streets into the solace of the St Regis Hotel. There he rang down from his nineteenth-floor, pistachio-green suite, where he was reading the *New Yorker*, to Rex on the fifth, to say that he was too tired to dine with Tallulah. He was saving his energy for Willa Cather.

Ever since he was a teenager, Stephen had been fascinated by Cather, whose novels evoked the sea of grass of Nebraska and the prairies where she had grown up as a tomboy, complete with a crewcut. Now she lived in New York, where she would walk barefoot in

Central Park every day to reconnect herself to the land. After writing fan letters to Cather for nine years, Stephen was about to meet his heroine. She greeted him at the door of her Park Avenue apartment wearing a flamingo-pink satin tunic and black satin trousers – he couldn't believe he was talking to the author of *A Lost Lady* and *My Ántonia*. Over tea, these two gender-challenging people – an English aristocrat and an American pioneer – discussed Cather's favourite places in New England. And so, a few days later, Stephen left New York and Tallulah and Cole Porter and Elsa Maxwell behind and journeyed north to Massachusetts where he settled for the winter, like a migratory bird.

He checked into Mount Pleasant House, a shining, snowbound hotel in Jefferson, its verandah overlooking the same mountains that the Fruitlanders saw from their utopia. And there, in the New England countryside where the Apostles of the Newness had wandered penniless in chintz smocks, Stephen, an itinerant soul insulated by his own sense of beauty and letters of credit from London, watched as the snow fell in 'great white goosefeathers', their drifts casting 'the longest sapphire shadows'.

In his solitude he'd take the bus into the nearby town of Worcester, passing wooden houses which looked rather temporary, as if they were made of matchsticks, the rocking chairs on their porches put away for the winter. The countryside rolled past the bus window, its low pines and sandy grass all dusty with frost and snow. One evening Stephen walked to the Jefferson village post office as the sun was setting. 'The snow was an oblique toneless white, the sky startling in its clarity … shaded like some tropical bird, soft saffron yellow – deepening to flamingo pink & then to violet … the cold seemed to hold & embalm the moment.' Stephen had only stepped into the shop – to buy more amethyst-coloured vases – when the sky had turned 'gentian blue, the stars in freezing splintering splendour – the snow a dim miraculous glimmer on the ground'. He felt blessed by the place, free in this New England.

On Christmas Eve, Stephen left Jefferson for Jaffrey, being driven thirty miles north – *Pas plus vite, monsieur, nous ne sommes pas presses* – into New Hampshire. The Shattuck Inn was still run by the same family, the Austermanns, whom Cather had known when she had begun her visits there twenty years before, writing her novels in a tent in a field before moving into the inn, and rooms which were reserved for her at the top of the building so she wouldn't be disturbed by other guests walking overhead.

Like Thoreau, Stephen wrote in his journal because he was lonely. He liked the Indian names of the places around him – Pawtucket and Lackawanna – and relished the fact that he was only an hour from the house in which Emily Dickinson had spent most of her life as a recluse – 'one hour from Paradise in any latitude'. All the while he received a stream of letters and telegrams from Cather, anxious that he should be enjoying the place she loved: 'Please write to me when you can, everything you think about the country, and everything you do there will be of intense interest to me.'

The Austermanns were certainly solicitous of their titled guest. Their son George – 'such a nice boy . . . I was reminded how well I like ordinary brown hair' – would drive Stephen into town so he could shop for snappy American clothes and admire the Christmas lights, so much more gay and splendid than those in England. A handsome Norwegian ski instructor, John Knudson, took Stephen to a trail in the woods which he was working on. Stephen sat on a pile of logs listening to the chickadees as John chopped down saplings; he even cut down one young birch himself, '& only felt sorry 3 hours afterwards – we were so happy & busy'. John told Stephen about his brother in the Canadian Mounted Police, '& we discussed the joys of camp life'.

Stephen's love of nature sat oddly with his life of artifice. He revelled in pigeons and chipmunks and mice – 'when I want solace I think of the raccoons faces'. But like Thoreau he surveyed his utopia of one, created in his own image; there was no room for anyone else. And where Thoreau was deemed ugly and Stephen beautiful, and the former lived in a wooden hut and the latter in a stately manor, the two men's spirits shared the same state of separation and ecstasy, standing apart from the world in order to better regard it. They even shared the same poetic disease, consumption. Perhaps they might have set up home together, these two, holed up in another hut in the woods, albeit one exquisitely decorated, probably in chintz. If Stephen could win round Willa Cather or Virginia Woolf, I daresay the philosopher would have soon been rapt by his tales of curious mice and the latest haute couture.

But that winter, New England outgrew its romantic appeal. Stephen was haunted by his broken relationship with Sassoon – and despite the comforts of Whipple's Cosy Café, of trinkets and little vases bought in Jaffrey's shops, of excursions to Boston in the snow, drinking sickly liqueurs in smart hotels – in spite of all of this, he fell ill, as much with nostalgia as anything else, and not even Cather could console him. 'You have seen our winter at its most terrible,' she wrote; 'there has been nothing like it since 1917. But there's a kind of glory about blizzards, don't you think? not when you are ill though, poor boy.'

Extreme weather seemed to follow him that year. Where he had been disturbed by the storm at Seaford and its rising waves, Stephen now felt oppressed by the snow that came dun-coloured with sand from the west, besmirching its beauty. His mood lowered as the season dragged on, and the glamour was replaced by a greater grimness. The gay nineteen-twenties in which Stephen had shone like a shooting star had been replaced by a different, darker, more punitive decade.

'Europe continues to grind its teeth,' he wrote on 11 March 1936. 'Hitler has re-armed the Rhine, France is hysterical with fear – rushing guns & planes to the area. Flaudin demands that German troops be ousted. Eden wishes to consider Hitler's peace pact suggestions . . . Nazi troops by radio-wireless photos entering Cologne – while girls cheer & pile flowers on to them. Outside my window – old dirty snow slips with a soft rushing sound off the roof – mild rain – wet air blows in.' Stephen's response to European politics was to order perfumes from Paris via Boston – Caron's *Fleurs de Rocaille*, Millot's *Crêpe de Chine* (which Siegfried had once given him in the Pavillon Henri IV, St Germain), Guerlain's *Shalimar* and *Liu*, Chanel's *No. 5*, and Gabilla's *La Violette* – an exotic, narcotic library whose old-world scents might, by their civilised fragrance, exorcise the evil stench.

But these evocative distillations – many of them containing ambergris, the essence of the ocean derived from deep-diving whales once hunted from New England's whaling ports – failed to achieve their magic. Here, on 14 March 1936, Stephen's journal ends abruptly with a pencilled note, in what I thought was a literary quotation – 'thinking of the wonder & exstasy of being alive – I feel that although I am often unhappy & disappointed – it is never life's fault – always my own ...' – yet which, when entered into a search engine, comes up with references to clinical depression.

Twelve years later, after the world had been overturned by war, Stephen was taken to a clinic outside London and electric currents were passed through his brain. The rubber contacts were placed on his golden hair, as they would be clamped onto Sylvia Plath's blonde head, the Bakelite switch was turned on, and Stephen's body shook with convulsions. The machine sought to shock him out of a melancholia that a later tenant on the Wilsford estate, Vidia Naipaul, would diagnose in his reclusive landlord as accidia or 'monk's torpor'. It left Stephen bereft and abandoned.

The violence – external and internal – had even threatened his home. During the war Wilsford had been requisitioned as a military hospital; the nurses complained that wounded soldiers should not be expected to use lavatories which had been painted gold. For his part, Stephen's greater contribution to the conflict had been to climb one of the Wiltshire downs and lay a bouquet of flowers at its summit. He was right, of course: chalk turf and rising larks were more sacred than any cenotaph. But Stephen closed his eyes to death, and when Virginia filled her pockets with stones and stepped into the river to drown herself, he could barely bring himself to register the loss.

He had spent most of the war in a suburban-looking nineteen-thirties flat in the centre of Bournemouth. From there he wandered the parks and chines whose pine-scented air was believed to aid the tubercular, as well as evoking his beloved Mediterranean. His fellow consumptives Robert Louis Stevenson and Aubrey Beardsley had come here for their health, although the resort failed to expel the infection from their lungs – Beardsley had haemorrhaged onto the clifftop path, leaving vivid crimson flowers in his wake. For Stephen, walking or bathing on the wartime strand was problematic: it was strung with barbed wire to keep out the expected invasion.

With the coming of a sort of peace, Stephen returned to Wilsford. As the trees grew taller around the silent manor, he retreated into past glories and future triumphs. He became obsessed with his would-be masterpiece, *Lascar: A Story of the Maritime Boulevard*, whose quayside tarts and muscled Lascars he drew over and over again in increasingly crowded images, as if they couldn't be confined by the page: the women extravagantly costumed like operetta versions of themselves, the sailors' bodies graphic with tattoos, while their faces all seemed to have the same features, the idealised matelot of his dreams. Spilling out into its margins, Stephen wrote endless notes to himself in ever-changing ink, cerise and turquoise and vermillion, along with poems about a boy with broken rose-red wings falling into a sea full of tears. He announced, in a self-published flyer printed in Portsmouth, that his book's spiritual godparents were Joseph Conrad, Emily Brontë and Herman Melville, that it would bear 'the sea's illimitable detachment' and 'restlessness and mysterious moral strength', and he had its cover published in *Horizon*.

But his work would never be imprisoned between hard covers. Like Orlando, who produces screeds of plays and poetry, there was a sense that it would be vulgar for Stephen's work to appear in public: 'to write, much more to publish, was, he knew, for a nobleman an inexpiable disgrace'. As the modern world encroached and he went nowhere, he entered an extended reverie of retrieval, travelling backwards into a vanished life preserved in his journals,

which now lay gathering dust in his house. 'It was dawn – I was flying ... I remember the seaplane as it rose above Southampton – slowly turning – and then going on straight – on & on into the first delicate radiance of early sunrise.' In his mind's eye he was still sailing down the same waterway in an ocean liner with a retinue that included his valet, his nanny and his plush monkey; or taking off from another airport: 'The Prince of Wales is getting into his aeroplane which stands next to mine.' He was lunching with Gertrude Stein in Paris, and taking tea with Willa in New York, or checking into Brown's Hotel in London: 'This morning when I ordered breakfast by telephone the waiter said, "Yes, Madame."'

And just as he was forever rewriting *Lascar* in kaleidoscopic colours – merely starting back at the beginning once he had reached the end, so that the entire enterprise became one endless loop – his house was continually reset as a stage for his memories. Its many rooms, once so full of family and visitors, like Mrs Ramsay's, were now filled with things rather than people. Straw hats from lost summers stayed on the newel posts of the stairs. Stephen replaced emptiness with dreams. His letters lay piled by his side, his jewellery and make-up and books and scarves gathered about him. The pale tide of a pure white carpet laid down when Wilsford had been a luxurious heaven in the nineteen-thirties, like some private film set, was now splattered with starbursts of carmine and ultramarine and emerald as Stephen dipped and discharged his pen, leaving shiny mineral-animal deposits like ground-up shells or crushed butterflies. Outside in his English garden, he ordered twenty-two tons of silver sand to be spread on the lush green riverbank, creating a simulacrum of a Côte d'Azur beach, complete with wheeling palm trees. Tropical birds and lizards escaped from their glass houses and into his illuminated manuscripts; in the winter they took refuge in the conservatory, where they kept Stephen company, listening to *South Pacific* on his Dansette over and over again.

Meanwhile the sea invaded his bathroom, its tub and basin filled with shells on which Stephen left the taps running, since, as we all know, they look better that way, the way they do in the water. Stephen himself had been left behind by the tide, like Orlando, who returns to her palace of three hundred and sixty-five rooms surrounded by lawns like a smooth green ocean, wandering into a bedroom that 'shone like a shell that has lain at the bottom of the sea for centuries and has been crusted over and painted a million tints by the water; it was rose and yellow, green and sand-coloured. It was frail as a shell, as iridescent and as empty.'

It was now that Stephen's obsession began to overwhelm him, in waves of memories merging into one so that it was difficult to know what was real and what was make-believe, whether he was in Wiltshire or Acapulco. His house became inundated as surely as Brighton Aquarium in the storm of 1935. Multicoloured fishing nets from San Francisco were draped over the banisters as if to catch passing fish; the shells in the bathroom spilled out onto the landing and down the stairs; and *Lascar* became an ever more elaborate reef of words and images, encrusted with salt and ink.

To Stephen, as to Virginia, the sea was a place in which to be subsumed. As young students at the Slade, he and Rex Whistler had united in their desire for a Kingdom by the Sea, an image from Edgar Allan Poe's last, doomy poem, 'Annabel Lee', which they both loved: 'I was a child and she was a child | In this Kingdom by the sea.' At that peak of his youthful beauty, Stephen had posed for Cecil's camera on the rocky shore of Cap Ferrat clad only in a fake leopardskin, and had sketched himself half-drowned in a pool, a Narcissus under a serious moon. Thirty years later, Beaton captured Stephen in a photograph in which, purposefully propped up on an ormolu table with its cover turned to the lens, was a copy of Rachel Carson's *The Edge of the Sea*, a book devoted to 'the place of our dim ancestral beginnings'.

Stephen was now a spectacle all of his own, washed up with his treasures on his bijoux-strewn divan. Beaton brought Isherwood and Capote and Hockney to listen to tales of dear Morgan and Virginia's peculiarities; Kenneth Anger and Derek Jarman came to hear stories of Garbo and the Ballets Russes and planned to cast him in their films. Stephen's life had become a movie clip endlessly rerun, ever more scratched and misted.

And there, one afternoon in the autumn of 1986, I was admitted to his darkened bedroom in that closed-up, sleeping house. He lay on an ice-blue satin bed in a room with the curtains drawn tight against the day, as if he might fade away in its light. One bar of an electric heater glowed in the corner; a bare bulb shone faintly from the ceiling. The wind shook the windows and rattled at the latches, stirring up the leaves outside and pushing them through a gap at the foot of the front door. The ancient wilderness was ready to take over, only biding its time. Out of the gloom, Stephen offered me his hand, weighed down with a scarab ring. I reached out with mine. In a fluting voice from another era, he asked, 'Do people still think of me in London?' As we shook hands I was conscious of what was passing between us: a secret world, and all the people I had never met.

Then I left him to his dreams and wandered alone around the great empty house, its every surface, from the stairs to the floor to the furniture, covered with books and paintings and letters. Downstairs, in a silver-curtained library, draped over a silver satin chair, was Stephen's hussar's coat, its epaulettes and braid spilling over the back of the seat; there were Venetian silver-gilt grotto chairs carved in the shapes of shells, and around the room ran a frieze of plaster scallops, encircling his kingdom beneath the sea. All the while, in the stillness, I could hear Stephen upstairs, laughing and talking to himself, as if the party were still going on and I was the last guest to arrive.

One year later, the Great Storm of 1987 swept up the Channel and into southern England; its hurricane-force winds felled fifteen million trees. At Wilsford, the palms swayed in the garden and as the storm rose to its height that night, one enormous branch smashed into the glass conservatory.

But none of this disturbed Stephen, since he had become ashes, and his earthly remains now lay in the churchyard next door, slowly sinking into the soft green turf.

THEPEOPLEOFTHESEA

A great white block rises out of the water like an iceberg. From this distance its top looks blurred, somehow in motion where it meets the sky. We are far from seeing the cause of this animation before we smell it: an intense aroma of ammonia; the sea, digested. Only as we draw nearer does the scale of the rock become apparent – along with the source of the stink.

The island is three hundred feet high, and circling its summit are gannets: thousands of them. The air cracks with their cries, celebrating their ownership of this place. Bass Rock is a plug of volcanic basalt left behind when the North Sea washed softer rocks away. But it isn't black: it is iced white with the birds' bodies and their guano, a wedding cake dropped into the cold waters. It is total bird world: voraciously, noisily, all-engulfingly bird; a Haitian bidonville, a Brazilian favela of birds, as if the volcano had burst open and is spewing birds from the centre of the earth.

The sky darkens, the light diminished by wings as wide as a man is high. No seabird is anchored to land; only its shores or its islands mean anything to them. Here, barely half an hour's train ride from Edinburgh with its elegant terraces and lofty castle, the world's largest colony of northern gannets – as many as a quarter of a million birds – perch on ledges or whirr around the rock like a cloud of gigantic gnats. Densely concentrated at the centre and spinning out in a vortex, their flight paths continually criss-cross each other, guided by some unknown traffic controller, each in their helical cycle spiralling from the eye of the hurricane to its outer edges, where the swirling whirlwind breaks up as birds plummet into the water at breakneck speed, sixty miles an hour. Where whales fall upwards out of their world, gannets dive down into it; the one for the other. It is a mission for which they are supremely suited, these fish-seeking missiles, with their forward vision and slit nostrils, their bills strengthened with bony plates and fitted with inflatable sacs to absorb the impact like a car's airbags.

I'm so excited that I perch precariously on the rubbery side of the boat, and have to be reprimanded by the skipper. I crane my unreinforced neck, trying to reach up into the sight. The sun haloes the summit as the birds create their own aurora; there's more gannet than sky. They occupy three dimensions, like a school of fish; a single sensate mass echoing the sea, eddying with its currents. Then, as we move into the lee of the Bass, we come face to face with the raucous heaving throng. There are gannets on every possible ledge. The down from the gugas – puffball chicks whose new coats make them look bigger than their parents – drifts dreamily in the air. We're sailing through avian snow.

Every now and again juvenile birds launch themselves from the lower rocks. A gannet's narrow wings seem stiff, almost fragile; there's more than a little of the albatross about them, and they dip down towards the waves before they gain enough lift to join the others. Once there, their endurance is formidable; some fly as far as the coast of Norway to feed. It is impossible to look at one bird in so many, to discern the single defining shape that makes up so many others, constantly forming and re-forming over my head. If the cormorants' outstretched wings signify the holy cross, then the black-tipped gannets resemble flying crucifixes, emblematical missionaries gliding in the air, familiars of Baldred, the Dark Age anchorite who lived on Bass Rock, walled up in a cell while the birds flew all around him.

As we round the rock – itself riven by a fissure that runs right through it, as though it had been stabbed – the lighthouse comes into view. Short and stubby, built on a platform to withstand the weather, it has been appropriated by the birds; it's even painted in the same colours, the pure white of their bodies and the blushed ochre of their heads. The ruins of a medieval chapel have long since been subsumed, as has the castle that held Jacobite prisoners. The gannets have so completely colonised the rock that they even take their name from it – *Morus bassanus* – or is it the other way around? Nor are they content with their occupation: each blue eye stares out its neighbour as they dispute every inch of territory. Their dissent seems human – individual gannets may live as long as we do – but they predate us by twenty million years. Below them, arrayed on the lower rocks are their dark cousins, cormorants and shags. The latter glitter green with emerald eyes, necks subtly arched as they stand there, carved out of basalt, marquises to the cormorants' earls.

All these animals couldn't be more of the sea unless they were fishes themselves. They are the extraordinary that marks the ordinary; the focus, the mass, all pulling together like the encircling waves and tides. There is no human dominion here, although it is respite from the hunters who once shot them for their food and feathers that has allowed them to build their bird city. And as we pull back towards the shore, I realise there's another audience for this overwhelming scene. From the waters of the facing cove, under a hill topped by a whale bone arch, pairs of big black eyes peer over the long whiskery snouts of grey seals, taking it all in.

In the late nineteen-forties, as Stephen Tennant was retreating into his Wiltshire estate, the writer and radio producer David Thomson set off on a series of journeys to Scotland, Ireland and their islands to gather stories of selkies, spirit seals – mythical creatures, part human, part pinniped. His book, *The People of the Sea*, published in 1954, is a lyrical record of beliefs held by coastal dwellers, already half-lost in the second half of the twentieth century. It is told in the voices of the people of the sea, sometimes as though the selkies themselves were speaking, their mournful voices drifting restlessly across firths and loughs. Thomson recounts, as if he'd just walked in on the conversation, one story told to him by an Irish teacher. 'And I remember how they used to say that anyone that's born of the water is not able to sleep. 'Tis like as if the movement of the water was always with them in their head.'

His informant, who appears as a dramatic character as real as the writer allows him to be, goes on. 'With the hare and the rabbit and the rat and the mouse and all the creatures of the land you'll find they so waken and go asleep the same as we do. But for them that has the tides of the sea running through them from the day they are born until the day they die there is no such thing as sleep the way we know it.'

Thomson, whose family had aristocratic links and who was the son of an Indian Army officer wounded in the First World War, was brought up in his grandmother's 'sea-lit' house on the outskirts of Nairn, on the coast of north-east Scotland; he had been sent there partly because at the age of eleven his eyesight was damaged in an accident, marring his vision for life. Sound became his sight, his most important sense; he was always listening, intently. After working as a tutor in Ireland in the nineteen-thirties, he joined the

BBC in 1943, making programmes on the Irish Famine and animal folklore. In the pioneering spirit of Mass Observation and the film documentaries of the time, he investigated the selkie, caught between human and natural history. As a boy he had thought of wolves and seals, and associated the death of a seal with the death of the albatross in Coleridge's poem, which he learned by heart at school. In his adult writing he attained 'a state of almost animal consciousness', as Seamus Heaney observed. Rather than anthropomorphic, he was theriomorphic, merging into the creature he was describing. Sheltering in a barn, he becomes part of his surroundings: 'I heard a raven croak twice. I felt the autumn coldly on my face, but because this old cowshed had been lately used for dipping sheep there was a smell of dung as though the warm life of the farm lingered on.' Heaney saw Thomson's stories as shadowed by nuclear war; elegies for an age that progress was about to destroy.

Thomson's book begins in Nairn, 'a hard, small town' caught between mountains and sea, 'which on fine days lies opposite the blue cliffs of Cromarty and on grey days looks out at a rigid black skyline, very close and broken in the middle by a gap called Cromarty Firth'. These are sharp places, holding out. They're not soft or comfortable. They bear the relics of war: Cromarty's cliffs are still strung with rusting gantries from the First World War, wrapped around the rocks like metal wrack. Further up the inlet, facing the Black Isle, decommissioned oil rigs rise out of the water on corroding pillars like drowned temples. Houses cling as limpets to the land; low block-like cottages run sideways to the sea rather than face it head on. Settlements peter out by the shore, as if they'd ceded defeat, seeking a truce with the sea.

Here people are part of, rather than apart from, the water. The herring fishery pursued its silver darlings along these coasts; my great-grandfather, arriving from Ireland in the eighteen-eighties, would work in the same industry down at Whitby, where his wife may have joined him in the trade. Fisherwomen were known as herring quines, as Elspeth Probyn notes, so covered in fish scales as they prepared the catch that they looked like fish themselves. In 1859, Charles Richard Weld, historian to the Royal Society and organiser of John Franklin's polar explorations, observed on a visit to northernmost Wick that the women 'all wore strange-shaped canvas garments, so bespattered with blood and the entrails and scales of fish, as to cause them to resemble animals of the ichthyological kingdom, recently divested of their skin, undergoing perhaps one of the those transitions set forth in Mr Darwin's speculative book ... If a man may become a monkey, or has been a whale, why should not a Caithness damsel become a herring?'

And in a strange way, Weld's vision melds with an older Scottish one, that of the mysterious animal which the tattooed Picts carved on their stones and perhaps inked on their limbs, along with all their animal-headed human hybrids. The Pictish Beast might be merely fantastic, a water kelpie or a loch monster; or, as some contemporary historians have speculated, an echo of the bottlenose dolphins that swim in Scottish waters, or even a rare beaked whale. But to me it looks like we all do in the womb: the beginning of all animals, a curling, snouted creature rearing up on unfurling fins in some extended process of evolution.

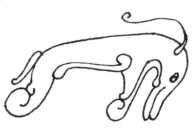

I can't imagine I'd ever feel at home in this place, for all my father's love of Scotland and our childhood visits here. The North Sea roars along wild beaches; I could never befriend these waves. They are not built on a human scale; they are untamed, barbaric and brutal. These are not resorts. There's no balm here, only the offer of more injury. Their sentinel lighthouses, as constructed by the Stevenson family, admit as much: 'Battered by storms, ravaged by waves, built by Mr Stevenson.'

In feats of extraordinary engineering, the Stevensons sought to defy the sea. On Bell Rock from 1807 to 1810, Robert Stevenson built a 115-foot-high lighthouse, one of a dozen he would create around the Scottish coast, and commissioned Turner to commemorate man's mastery of the ocean and, perhaps, its animals: when Robert took Sir Walter Scott to visit the site, the celebrated author noted the presence of 'several seals, which we might have shot, but, in the doubtful circumstances of the landing, we did not care to bring guns'. From 1838 to 1844, Robert's son Alan struggled to erect Scotland's tallest lighthouse on Skerryvore, a remote reef off the Hebrides. He used three hundred charges of dynamite to excavate a space in the rock; the 156-foot-high tower he built was said by his nephew Robert Louis Stevenson to be 'the noblest of all extant deep-sea lights', sweeping upwards, slender and curved.

These man-made responses were both ingenious and futile. In 1849 Alan Stevenson constructed the evocatively named Covesea Skerries light off the coast near Nairn; he was followed in the family tradition by his brothers David and Thomas, father to Robert Louis. Between them they built Whalsay Skerries light in 1854, and more than thirty others. But unlike his siblings, Thomas Stevenson was fascinated by the *qualities* of his appointed adversary. He became obsessed with waves: the forces that determined what he built, and which haunted his dreams.

Thomas appeared to see the power of waves as supernatural; his son saw a certain melancholy in his father. Thomas measured them, assessed their height and length and volume using a wave pole and a dynamometer, and asked lighthouse keepers to record wind speed, height of spray and water pressure per square foot. The sea had become a laboratory for his ideas. He even tried to weigh waves, and published his findings in the prestigious journal of the Royal Society of Edinburgh. But 'his studies of waves were only descriptive', as Rosalind Williams writes. 'He was not able to reduce their tumult to a scientific understanding robust enough to make predictions.' Quite simply, the sea was beyond him and beyond science. Waves were proof only of God.

That same sense of tension and intangibility – and morbidity – would suffuse the work of Robert Louis Stevenson, who replaced the technical concerns which had caused his father so much angst with metaphysical ones. Down the coast at North Berwick, off whose harbour walls I dive along with the local children, Robert looked out to Bass Rock, whose lighthouse would be built by his cousin David; his Treasure Island may have been suggested by the islands of the Forth. Stevenson was an admirer of Melville, and was admired in turn by Jack London, who quoted him in *Martin Eden* and visited his grave in Samoa. It was the wide, wide sea, from Scotland to the South Pacific, that supplied Stevenson with his inspiration.

As a young man accompanying his father to a construction site, Robert put on a diving suit and, encased in brass and rubber-ised canvas with lead boots on his feet, he was lowered below the waves. 'I was like a man fallen in a catalepsy,' he wrote, as he de-scended into the twilight of the sea. 'Looking up, I saw a low green heaven mottled with vanishing bells of white; looking around . . . nothing but a green gloaming, somewhat opaque but very restful and delicious.' As he returned to the surface, 'I shot at once into a glory of rosy, almost of sanguine light – the multitudinous seas incarnadined, the heaven above a vault of crimson.' It was a kind of rebirth, Billy Budd's fate reversed. In *The Strange Case of Dr Jekyll and Mr Hyde*, one of his few stories not set on the sea, written in 1885, around the same time Melville was writing about his Handsome Sailor, Stevenson describes another descent: that of a good doctor into an evil man in a manner which recalls his father's crises over the meaning of waves, and Darwin's observations on the descent of humans. For Dr Jekyll, the 'agonized womb of consciousness' is the 'curse of mankind'; it only underlines 'the trembling immateriality, the mist-like transience, of this seemingly so solid body in which we walk attired'.

Stevenson wrote his book, not in the Edinburgh streets that had inspired it or North Berwick's rocky islands, but in his Bournemouth villa, built on calmer, sandy cliffs, and which he named Skerryvore. Suffering from the consumption that had prompted his southerly migration, he found he could not go outside: he was often too ill to leave the house, or too dogged by fans of his work who clamoured around him, to step out and take the air which had brought him there. Like Joseph Conrad, Stevenson saw darkness and domination reflected in the imperial sea. He would escape it for the South Seas and their clear, warm waters, far from all the constructions built by his family and their ever-burning, ever-turning lights, warning of unseen obstacles and the perils of leaving the land.

As the night gathered behind a full moon over Spey Bay, I went out for a last look at the water in which I had swum three times during the day. Where the mountains gave way to the sea, the waves thundered in, as if to pick up the land and put it in its place. In the darkness, the backlit clouds above were a ghostly mimic of the scene below. It was terrifying and sublime: the sheer force of the North Sea only a few feet away, and yet almost invisible. Had I closed that distance, as I had during the calm light of the day, I would not have lasted long.

This is a raw, unformed place, still in the process of sorting itself out. It seems to rumble with its becoming, just as the huge cobbles on the beach roll under my feet like ball bearings, threatening to tip me this way and that as I dodge huge tree roots and other giant pieces of flotsam. The oversized scale reminds of an antipodean wilderness, of New Zealand's South Island or Tasmania's south-western shore, as if I'd been helicoptered into nowhere. There's no accommodation of the human: no access, no mediation, no agreement. It's all in dispute, with itself.

Seventeenth-century stone ice houses, dug into the ground like air-raid shelters to preserve seasonal gluts of salmon, lie under turf waves. The Spey spews three-foot fish into the mouths of bottlenose dolphins waiting like grizzly bears on a salmon run – grown so rich and fat on the supply of protein that they're among the largest representatives of their species in the world. While they leap out of the water within yards of the shore, their sedate seal cousins lounge on the rocks, subtle in their various shades of grey, beige, taupe, charcoal, stone, like some fashionable paint manufacturer's chart. Their eyes implore a semi-human state; as if they were humans in the process of becoming or unbecoming. Arrayed on a far beach across Findhorn Bay, under a sandy scarp topped by dense dark pines that might mark a remote strand in British Columbia, their animal nature is undermined by those eyes. 'There were little sandy beaches where no one had been since the end of time,' Mr Ramsay says in To the Lighthouse. 'The seals sat up and looked at you.'

These are accessible marine mammals, all too easily hunted, hauling out on the land, hugging it with their languorous furry coats, almost modelling their allure. For centuries they have provided food and oil from their meat and blubber, and as David Thomson relates, their skins supplied clothes and caps and purses sleek with their thick pelages, still alive with the sea. A purse made from seal fur was said to react to the weather, its pile rising like a hairy barometer –

'Aye. It's sleekit now. The hairs is lying even down like the hairs of a sleepy cat. But in the morning likely they'll be standing like the bristles of a sow.'

– as if even in death the animal had a life of its own.

'Did ye ever hear tell of another skin that would live that way, after one hundred years?'

Some fishermen kept scraps of sealskin as charms against drowning, like cauls. When I was a boy, a friend returning from Norway gave me a little seal made of its own fur, with bead eyes and nose. I loved its glossy hard pelt, both soft and stiff at the same time; it made me think of stories of Romany children from the forest whose mothers sewed them into rabbitskins for the winter, like wild inhabitants of the woods. For the people of the sea, the seal too merged between myth and reality: between the selkies – who came onto land in the guise of a human and kidnapped a lover, grasping them in an inescapable embrace, breathing deeply into their mouth before diving off a cliff and taking them down, down deep into their world – and the animals that supplied them with food and warmth and light as they burned their oil in lamps, yet which they killed as predators of the fish to which they, the greedy humans, also laid claim.

If seals look and sound human, it's easy to imagine that they might become us, or we become them. I've often shared northern shores with these animals. In the Outer Hebrides – the Islands of the Strangers where Britain fractures into the Atlantic and the landscape looks so alien, with its puddle-like lochans dropped like lakes from the sky into burnt-orange peat bog, that it stood in for the infinite beyond Jupiter in 2001 – I've slipped my skin to swim in surging white and turquoise waves as grey seals hung in the surf, their limber shapes against the shot-through light, watching me curiously. Across the ocean off Cape Cod, their peers spy on me through the same blue water. Even in the Solent, seals have popped up next to me, surprising me as much as I surprise them. They are sentinels of the near shore. Their heads bob in the surf like buoys, or they weave about below my legs, their dexterity far exceeding mine. They follow as you walk down the beach, accustomed to anglers casting their lines; a fishing rod signals supper to them. For all their physicality, they're caught between one element and the other, as Thomson's book would record.

'But I suppose now,' said Tadhg, 'that the creatures of the water, not being able to sleep, I suppose that they wouldn't be tormented in themselves the way you and I would be lying awake at night.'

'I don't know then,' said Sean. 'But them that was that way enchanted in human form suffered everything that you or I would suffer if we lay awake for years. And worse. Or so the old people used to tell me. And the seal above the others, Tadhg, you know well, for they say she is something differing from every creature.'

'But what is in the mind of them, I don't know – the creatures,' says another islander; he too sees the seal as a symbol of the mortality of the sea. 'She went out with the ebb tide,' says Michael the ferryman of his wife, two months deceased, 'the way I knew she would, for I was looking from the window when God took her and I never saw the water lower than it was that minute. I thought to myself, and I still praying, if God spares her now for those few minutes, and the tide to turn, she will be safe.'

I remember my parents' gaunt faces as, separated by a decade, they lay back on their respective beds, the tide gone out of them, their cheekbones as high as rocks on the shore. The end of the land is the end of all this. As Molloy says, from here on you can only get wet, or drown.

Does death always come at low tide? 'No,' says Michael. 'But if a beast or person is anyways weak, then their strength will fail and build with the ebb and flow. 'Tis just the same with the moon, though you wouldn't remark it in yourself, your mind being strong; if you watch a person that's weak in the mind you will see how his sense comes and goes with the moon.'

Thomson's tales of the people of the sea turn from one story to another, declining to decide between myth and reality. As he wanders around the relic edges of the Celtic islands in the late nineteen-forties, a lost romantic in a black-and-white world, he listens to stories from another age: 'I came only just in time to hear the last remnants of pre-Christian culture.' These were places where sentinel stones still stood in the watery land, buried tide-high in the bog like whale teeth in their jawbones, placed there to tell out the phases of the moon, so much more important than the sun to people of the sea. A man whose sight was impaired spoke to people of the sea who still believed in second sight, especially when it came to tales of the other people from the sea. He hears of a baby accidentally left in a cave by the sea, thought drowned, only to be found being suckled by a seal: the child grew up to be 'a great man, a fine young man, and a remarkable swimmer. He was a great swimmer.' The fact is emphasised. Seal-hunters are reproved by the grieving voices of selkie-seals, plaintively calling,

'Who killed Anna? Who killed Anna? Who killed Anna?'

To be answered by another,

'Oh, the same man, the man always, the man always, the man always.'

On the coasts and islands, from the Outer Hebrides to southern Ireland, Thomson learns of the King of the Seals, chosen by a thousand million seals, rising up out of the waves, the size of a cow with the face of an old man, with limpets and periwinkles growing on his cheeks and head. Skinned seals lament their lost city beneath the waves, condemned to live on land, as hairless as humans. Men take selkie wives, hiding their pelts until the women, now mothers, discover the hiding places and steal back to the sea, returning only to leave gifts of fish and beautiful shells for their abandoned children. Neglected human wives turn to selkie partners for comfort, their children born with webbed fingers and toes, the webs successively clipped away by midwives and mothers and regrowing as horny patches on their hands and feet – 'the fins couldna grow their natural way, so they turned into this horn', like the toughened lumps on my hands, caused by my own seaborne condition that curls my fingers and toes as if to fuse them into fins (and which, for one Hebridean I meet, his palms a network of scars, prompted him to have his little finger amputated 'because it got in the way').

'For the seals and ourselves were thrown together in our way o' getting a living, and everything we feel, they feel, ye may be sure o' that.'

'I've heard a Shetland man say they are some o' the fallen angels. God threw them out,' says one islander; another that 'There was supposed to be a creature in the water for every one on land,' or, 'There was many a man drowned after seeing one.'

This world fades even as Thomson writes about it, like a distant recording. Yet his stories are anchored by the details and textures that define the hard edges of the land where it becomes the sea. The tar-covered curraghs like coracles which turn men into turtles, paddling aquatic creatures; and the way the fishermen dress as counterparts of the herring quines, becoming what they hunt or the land provides, their trousers 'woven at home out of wool, undyed in the weft and blue in the warp, which gives them a shadowy worn appearance. The fly buttons are not covered and the narrow trouser legs are split at the ankles, like cuffs. They wear heavy jerseys, grey or blue, knitted in an intricate pattern, and a sleeveless tweed waistcoat. Some have a serge-blue jerkin and some still wear the bawneen – an undyed jerkin almost white, for which they were so famous long ago. On their feet they wear pampooties – flatsoled shoes of uncured cowhide with the hair outside like the rivlins of the North. These grow hard and stiff if a man walks on dry land too long, but as the Aran Islander spends half his life in the sea, there is little danger of that.'

Here rituals, too, are holdfasts, like the late-night prayers over the dying embers of a fire, trusting to God that it will revive the next day.

'We rake this fire as the pure Christ rakes all. Mary at its foot and Brigid at its top. The Eight highest angels in the City of Graces preserving this house and the people till day.'

And so Thomson wanders off, out of his story, not caring whether what these fishers and crofters told him was true or not. That was not the point. 'As to the seals themselves,' he concluded many years later, looking back on his travels as a young man, 'no scientific study can dissolve their mystery.' Like the waves which frustrated Thomas Stevenson, the seals defied definition. To David Thomson, no other animal had such a dream-like effect on the human mind, or such a place in our unconscious. The selkie was part of our ancient narrative.

'But what is in the mind of them, I don't know – the creatures.'

The shape hangs in the water, a downed barrage balloon. Barely ten feet from the tideline, it resolves itself, with a rising snort, into a seal, suspended there, the same animal I first saw here years ago. Its black-grey back is ingrown with green algae, like those mythic whales said to grow trees and shrubs on their backs, as if they were another country. Kicking off my sandals, I wade out to it, worried that it's ill, lying there, puffing away like an old man with emphysema.

I'm right next to it now, alongside its bulk; I want to reach out and pat it and tell it things will be all right. Its whiskers are thick and yellowy and stubby and spare. It looks at me with that sidelong glance that dogs have, and which makes you vaguely concerned for what they might be thinking.

It's disturbing to be so close to such a huge animal on a suburban shore. It is longer than I am tall, and at least three hundred pounds heavier; an overweight surfer in a wrinkly wetsuit, or a hippo waylaid from some African watering hole. Its bones must be old and weary. The water is warm around it. I feel I could float there with it, basking, growing algae on my own back. It's just an old old creature, I realise. An old old seal, with more of the walrus about him. Best left alone. Later I see him further out to sea, still hanging there, still waiting. For what, I don't know.

He was there again this morning, in the dawn, gently lolling in the swell. I wondered if he recognised me, with that sidelong flash of his eyes, this King of the Seals, 'the biggest seal I ever saw, as big as any cow'.

'I remember well his face,' said Sean Patrick, 'for 'twas like the face of an old man.'

———

The mountains of North Wales seem built to keep out invaders. They hold the world at bay, as much a barrier as the sea. Mountains scare me. They make me think they're hiding something; that they're a dam against an unseen ocean, welling up with whales and seals and sharks and squid held behind the tides of deep time. These rocks turn the skyline barbaric, then fall sharply to the sea. Above me is Coed y Bleiddiau, the Wood of the Wolves, where, it is said, the last wolf in Wales was killed.

A century ago Porthmadog, set at the mouth of the Traeth Bach estuary, was a working port, then known as Portmadoc, exporting Welsh slate around the world. But it has long since fallen into disuse, and its harbour is full of recreational yachts rather than industrial barges. A restored railway carries tourists over the water and into the hills – the same toy-like trains I remember from holidays when cargoes of visitors were trundled past vast vertical fields of slate, tipped-up hills, as if the mountains were slowly slipping into the sea. I felt that I had only to take out one sliver and the whole lot would come tumbling down, Snowdonia and all.

Those silky slabs, originally formed as sedimentary layers of ancient seas in greys as subtle and various as the colours of sperm whales, and whose leaves occasionally fall apart to reveal fossilised creatures preserved like pressed flowers, were the saving of this place. This roof of Wales roofed the world. Portmadoc slate provided Netley's military hospital with its roof – the same shards I still find on my local beach, worn into rounded tokens, offered the drowned as their fare to the other world. At its peak Portmadoc was as busy as any pithead, a place of masts and men and machines. Ships arrived full of ballast to be exchanged for cut slate. In return, their rubble was dumped in the harbour, forming a foreign reef where Welsh trees are rooted in Mediterranean debris.

The town seems empty, an abandoned film set. When I ask for directions to my guest house, the man painting a shopfront claims never to have heard of the road, even though it is only three streets away. Up above, perched on the hill, a grand terrace stands testament to past wealth; its houses, too, might be sliding into the sea. On the far side of the bay lies the unreal Italianate village of Portmeirion, another import: a pastel-painted opera set overlooking the estuary where, that autumn, I attended a festival during which young people, freed by the lack of mobile reception, pulled off their clothes to dive in the water, and under cover of night watched performers with horned heads dance on stage.

This site was witness to other attempts at utopia. In the second decade of the nineteenth century, William Madocks, a Member of Parliament and property speculator after whom the port was named, employed a hundred labourers to build a great embankment across the estuary, draining it and creating new land. By damming the sea, Madocks sought to connect this Celtic shore with its counterpart, establishing a new trade route between Britain and Ireland – and from there to America.

This was the ambitious scheme that greeted Percy Bysshe Shelley when he arrived here in 1812 with his seventeen-year-old wife Harriet and a trunkful of seditious manuscripts. Still only twenty years old himself, he was a man on the run.

The poet had recently returned from Dublin, where in his attempts to distribute his *Address to the Irish People* he had been reduced to throwing his pamphlets from the balcony of his lodgings or running up to passing carriages and stuffing them into their windows. Although aristocratic by birth – he stood to inherit an income of six thousand pounds a year – he was a rebel and had been since he was at Eton, where he was known as 'mad Shelley' and had once surrounded himself with a circle of blue spirit flame in an attempt to raise the devil.

He was now doing something similar with his words. As his teenage wife wrote, a 'large box so full of inflammable matter' had been discovered by the authorities at the custom house in Holyhead. It was reported to the Home Office in London as containing 'a great quantity of Pamphlets and printed papers', as well 'an open letter, of a tendency so dangerous to the Government' that the Prime Minister was alerted. Shelley's demands drew on those of the French Revolution. He declared that 'Government has no rights' since it exhibited 'barefaced tyranny'; that 'No man has the right to monopolize more than he can enjoy'; and that 'Those who believe that Heaven is, what earth has been, a monopoly in the hands of the favoured few, would do well to reconsider their opinion.' Such rhetoric sounded more like a threat than a philosophy, one which would inspire Thoreau's civil disobedience.

At Oxford, Shelley had espoused atheism and free love, forswore meat and eschewed sugar since it was the product of slavery – ideas that would inspire New England's utopians a generation later, and which got him expelled from college. Now, in the eyes and ears of the authorities this hot-headed youth had become a dangerous activist, an anarchist who would call for the people to rise like lions after slumber. After leaving Ireland, Shelley had established his own commune at a lakeside house at Nantgwillt – the Wild Brook – in the Elan Valley, mid Wales. Each member had their own commune name, and would work the land and teach pupils from their library, turning the Wild Brook into a torrent of radicalism.

A year later, Shelley met John Frank Newton, who was a vegetarian on both moral and dietary grounds, citing the myth of Prometheus, the bringing of fire and the cooking of meat as the beginning of man's downfall. Newton had recently published *The Return to Nature: or a defence of the vegetable regimen*. In it he promoted a diet of vegetables and distilled water (pointing out that the Thames was full of animal oil, and that London's drinking water was contaminated with putrid animal matter). His children ran about naked in their Belgravia home, since their father considered clothes another bad habit of civilisation. The family ate only dried fruit and biscuits and drank weak tea at breakfast, with potatoes and vegetables in season for dinner, supplemented by macaroni and a dessert. Newton believed that man's 'chymical' state, from his dentition to his digestion, was 'wholly adapted to vegetable sustenance', and that diseases such as cancer were caused by eating flesh. He added that the 'monkies' of the menagerie in the Tower of London had sickened when given meat; and that the domestication of animals

... entails upon them many disorders and much misery. Sheep suffer in a way to call forth the most ordinary compassion ... After robbing the unfortunate creature of its own warm clothing, we keep it ready for the knife in a state of incipient rot, and then we exclaim, what a dull, sluggish stupid looking animal is this! I shudder at the thought which forces itself on my mind. Tell me, reader, is that originally noble creature man more, or is he less deteriorated than the mutton?

The suffering of animals echoed humankind's. Evoking Rousseau's noble savage, Newton claimed that the South Sea islanders were perfectly happy 'until Captain Cook conceived that they must be miserable without beef and mutton. He took compassion upon them, and they have since lost their former health.' He also observed that the whalers of Nantucket ate raw potatoes to ward off scurvy. His book took in new notions of the past, too, noting the discovery of a preserved mammoth in Siberia 'within a few paces of the shore of the frozen ocean', and citing George Cuvier's theories of extinction as proof that ancient myths were based on real events: the promised apocalypse had happened already. 'The burning of the world, which Platonic philosophers contemplated as being still in the womb of futurity, seems to have taken place long ago.'

Newton saw humanity in a state of perpetual dysfunction – 'It is not man we have before us, but the wreck of man' – and consumption and exploitation were the cause. It was a sentiment with which Shelley agreed, as would the Apostles of the Newness. But the poet's Welsh experiment, inspired by Newton and his family – they called Shelley 'Ariel' when he visited them, since he resembled 'the image of some heavenly spirit come down to earth by mistake' – was even more short-lived than Fruitlands or Thoreau's sojourn by Walden Pond.

Driven out of Wales, Shelley and his communists decamped to Devon. Occupying a cottage overlooking the sea at Lynmouth, the group aroused suspicion among the locals, with their unescorted females and a young man who was clearly 'someone' but they weren't sure who, and who exhibited distinctly strange behaviour. Shelley had overcome the problems of promoting his insurrection in a characteristically ingenious manner: he sealed copies of his *Declaration of Rights* in bottles which he called 'vessels of heavenly medicine' and set them adrift in the sea. They were talismanic, folkloric objects, like the gods-in-bottles assembled in the West Country and elsewhere, often by Irish workers: recycled glass bottles into which cryptic wooden items – ladders, tools and nails, apparently emblematic of the Crucifixion – were inserted, and which were then filled with water. In his ingenuity, Shelley also devised miniature boats composed of little boxes covered with bladders (clearly he did not object to this use of animals), with lead weights and sails to keep them on course; at least one was found by a fisherman, mystified by this suspect device.

Most ambitious of all was Shelley's sky-fleet of hot-air balloons, made of silk and cow gum and powered by spirit-soaked wicks; potent extensions of his prophetic visions of luminous airships and fleeting comets. Laden with his *Declaration of Rights*, they were launched like Chinese lanterns across the countryside, sailing over a landscape that was no longer a rural idyll but a battleground of fire, air and water. He hoped his missives would drift up the Bristol Channel, a conduit of slavers, offering 'a ray of courage to the oppressed and poor'. Had he but known it, his techniques would be taken up in the Cold War when West Berliners sent message-laden balloons over the Wall; they called them *Mauersegler*, 'wall sailors', another name for swifts.

Shelley's airships and boats might seem like toys to us, but Britain in 1812 was in uproar. It was caught in a war with revolutionary France that had caused the coast of southern England to be fortified, and fighting a new war with its former colony of America; its sovereign had been declared insane and replaced by a dilettante regent, the prime minister, Spencer Perceval, had been assassinated, and the country was threatened by Luddites determined to destroy the machines that were taking over their lives and livelihoods. In such an overheated atmosphere, Shelley's actions were an open challenge to an oppressive state. People were watching. 'Mr Shelley has been regarded with a suspicious Eye since he has been in Lynmouth,' wrote the town clerk, and when the poet's young and illiterate Irish servant Dan Healy was found guilty of 'dispersing papers printed without the printer's name being on them' and, rather than betray his master, spent six months in Barnstaple's grim stone gaol on Dartmoor, it was clear that Shelley and his revolutionists would have to move again. They left Devon at dead of night, on their own surreptitious flight.

Having reached the outer limits of the mainland in North Wales, Shelley believed he had found the resolution of his ideals: nothing less than a reformation of human and land. To the poet, Madocks's plan represented 'one of the noblest works of human power – it is an exhibition of human nature as it appears in its noblest state – benevolence – it saves, it does not destroy'. Like Ariel, Shelley was taking on the sea itself, subscribing to its circumscription.

In an echo of the Stevensons' attempts to conquer the waves, Madocks's Grand Scheme sought to embank the waters, reclaiming five thousand acres of land from the sea. Shelley did not see this as an act of theft. Rather, it was as if his own poetry might build the dyke; as if words would be used to tame nature and bend its fearful force, while reaching out to new lands. 'Yes! the unfruitful sea once rolled round these islands, through the perspective of these times, – behold famine driving millions even to madness.' Shelley had seen desperate hunger in Ireland and tyranny at work in England. Here in Wales he saw salvation in the sea. He pledged to spend his last shilling on 'this great, this glorious cause'. He even claimed that Madocks was the true Prince of Wales, ready to rule in revolutionary stead of the decadent regent.

Shelley's family and followers – Harriet, her sister Eliza, and later, when released from Barnstaple gaol, young Dan Healy (possibly tailed by the authorities, keen to discover where his seditious master had fled) – made camp in a sea-facing house, Tan-yr-allt, Under the Precipice – built by Madocks. Shelley called it 'a cottage extensive and tasty enough for the villa of an Italian prince'.

It was, and is, a sublime vista. At dawn, the sun rises over Snowdon, turning its crags pink. Below, the widening, fast-flowing estuary carries debris like meltwater from a glacier, eddying brown and swirling with silt. As I swim, unsure of what is deep or what is shallow, I feel the current is strong enough to carry me out to the Irish Sea. This place conspires: the entire inlet is unstable, shifting, defying – the very forces of Shelley's poetic imagination, his utopian intent.

But Madocks's efforts were not to everyone's liking; Harriet Shelley felt his actions to be against nature itself. 'The sea, which used to dash against the most beautiful grand rocks,' she wrote scathingly, 'was, to please his stupid vanity and to celebrate his name, turned from its course, and now we have for a fine bold sea, which there used to be, nothing but a sandy marsh uncultivated and ugly to the view.' Her husband, too, had ignored what lay below the ocean's skin, mistaking its open surface for emptiness; perhaps it was this disavowal that had stirred the demon which now stalked him in the Welsh hills.

That winter was intensely cold. 'The thermometer is twelve degrees below freezing,' Shelley told his friend Thomas Jefferson Hogg; 'this is Russian cold.' He felt frozen out by the people, too. 'The society in Wales is very stupid. They are all aristocrats and saints.' The poet retained his apartness. 'I continue vegetable; Harriet means to be slightly animal, until the arrival of spring. My health is much improved by it.'

But in February 1813, as the spring tides threatened the construction of the embankment and raised the tension between its champions, new storms blew in, fit to breach the barrier. Shelley felt under siege. A local landowner, the Honourable Robert Leeson – one of those aristocrats – had sent Shelley's seized pamphlets to the government; he may have been behind what happened next.

At the height of the storm, the pacifist poet took a pair of loaded pistols to bed. That night, shots were fired about the house. Windows were smashed, shouts heard and shadowy figures seen. In the confusion, Shelley found himself outside in the mud, with musketball holes in his nightgown. According to Harriet, who remained upstairs, terrified, the ghostly assailant shouted, 'By God, I will be revenged. I will murder your wife, and will ravish your sister! By God, I will be revenged!'

In these split perspectives, it is difficult to discern what happened, if anything happened at all. Thomas Love Peacock called it an hallucination conjured up by Shelley's overactive imagination and sense of persecution; but William Madocks declared it to be 'a transportable Offence, if discovered'. Others thought that the strange figure, a wild man of the woods which Shelley sketched on a firescreen like a gothic identikit, showed a man dressed up to look demonic, as though the poet were 'taunted and terrified by some deliberately contrived theatrical "apparition"'.

Years later, in an article in The Century magazine published in 1905, Margaret L. Croft investigated the incident, a detective coming upon the crime scene long after the event. She quoted a witness who said that Shelley had 'fancied he saw a man's face in the drawing-room window; he took his pistol and shot the glass to shivers and then bounced out on the grass, and there he saw leaning against a tree a ghost, or, as he said, the devil ... When I add that Mr Shelley set fire to the wood to burn the apparition (with some trouble they were saved) you may suppose that all was not right with him.' That vision, of a burning beech on a remote Welsh hillside, a raging fire to expel the devil, conflates with Shelley's retinal memory of a three-fingered and horned form, part stag, part demon, part tree, a kind of Caliban.

Ever since his boyhood, when he had sought ghosts in starlit woods, Shelley had feared a 'following figure'; a fear all the more powerful for the fact that he did not believe in spirits, gods or demons. Nonetheless, they haunted him relentlessly. The opposite was always true, in his contrary life. Whatever the truth of the feverish incident at Tan-yr-allt – whether it was a government-sponsored scare tactic, an assassination attempt, or a localised eruption of violent prejudice – Shelley would ever feel pursued by evil, even and until he succumbed to the waves.

———

He was, by all accounts, even those which stretch our credulity in the time and space between our selves and his, an extraordinary figure. 'Was it possible that this mild-looking, beardless boy could be the veritable monster at war with the world?' his friend Edward John Trelawny would write, recalling his first meeting with the poet. Tall but stooped, Shelley looked deceptively fragile. He wore expensive clothes, 'according to the most approved mode of the day', but they were unbrushed and rumpled. To Trelawny, he resembled a schoolboy who'd outgrown his uniform – albeit one who'd tried to raise the devil at Eton, drawing pentacles on the ground just as the starman drew cabbalistic signs on his carpet and saw demons in his pool.

The poet's face was pale and delicate and feminine and elusive – no one seems to agree which of the portraits are at all like – but his cheeks became freckled and tanned in the summer. His features were asymmetrical and powerful in animation. His eyes were blue and his hair was long and dark – or was it short and fair? – and he had a habit of running his fingers through its tangles. He was acutely aware of his body. At one point he believed he had contracted elephantiasis from an obese woman in a coach and kept pinching the skin on his hands, arms and neck, thinking his legs were about to swell to the size of a pachyderm's and his flesh crumple. He had no interest in food or drink: he appeared to subsist on tea, bread and butter, lemonade prepared from a powder in a box, and tincture of opium. He was prone to nervy gestures, as if he could not sit or stand still in the world, as if he were quivering in another dimension. Perhaps that was why he was drawn to streams and pools and seas; they stilled his querulous soul. 'Shelley never flourished far from water,' Trelawny wrote. 'When compelled to take up his quarters in a town, he every morning with the instinct that guides the water-birds, fled to the nearest lake, river, or seashore, and only returned to roost at night.' For someone whose life was predicated on extremes, the sea represented the ultimate, 'the Deep's untrammelled floor', where the Oceanids cast human fates, and where sharks gnawed at the bones of jettisoned slaves.

Shelley was destined for discontent. He had been expelled from Oxford as a REVOLUTIONIST after publishing a tract on *The Necessity of Atheism*. His 'strange and fantastic pranks' at college had included taking an infant from its mother on Magdalen Bridge, asking, 'Will your baby tell us anything about pre-existence, madam?' Thomas Jefferson Hogg, his college intimate, said he was once forced to stop Shelley inflicting violence on a boy who was maltreating a donkey; on the other hand, when he had gone out in a new blue coat with glittering buttons which were then torn off its skirts by a mastiff, Shelley went to fetch his pistols to shoot the animal. Hogg wrote of Shelley's sudden, pantomimic appearances, 'as he always looked, wild, intellectual, unearthly; like a spirit that had just descended from the sky; like a demon risen at that moment out of the ground'. He surrounded himself with 'queer people', as if to make his behaviour seem more sane; and as his imagination spilled into reality, his habit of scrawling and doodling on books extended to walls and wainscots and rocks, making his mark like a dog cocking its leg.

It is difficult to think of a more extreme existence for a poet. Around Shelley swirled storms and elopements, anarchy and abandonments, monsters and demons, suicides and premature deaths; events which could not but taint the tenor of his work and turn his life into gothic drama. Lives were lived in epistolic energy. The telegram had yet to be invented, but hurried notes might be delivered two or three or more times a day – between Shelley and Harriet and Mary; between Shelley and William Godwin, his father-in-law; between Mary and her father and her half-sisters Claire and Fanny; long declarations of passionate love or urgent demands for money – a sense of communication more fast-moving than ours because their thoughts seemed to precede the paper on which they were written.

Words were loosened, unbound, uttered on a semi-public stage in an age when philosophers and aristocrats kept open houses for ideas and conversations and gossip. Far from enlightening, real life for Shelley became so awful, with such evil returns for his attempts to do good, that he retreated into his poetry, which was more real than anything, and where he could achieve utopia, an equality of love and justice – for all that a poet of another age, Matthew Arnold, would see him as a 'beautiful and ineffectual angel, beating in the void his luminous wings in vain'. He was trapped by his times, like Walter Benjamin's angel of history, blown about by the storm of progress, his wings caught in the wind, no longer able to close them.

Shelley's story was so powerful and mad that it resonated beyond his work. In 1924, André Maurois's biography, entitled *Ariel*, would recast Shelley's life as a series of scenes from Maurois's own neurotic era. In the same spirit as Woolf's *Orlando*, Maurois equated the Regency with the Jazz Age, seeing Shelley as 'half man and half meteorite', his female followers as flappers or bluestockings, and his male friends as dandies and wits; all vile bodies in decline and fall. At one point, Maurois describes the effect of Shelley's Irish insurrection as a 'screaming joke'.

Bysshe sometimes sighed deeply.
'Is it necessary to read all that, Harriet dear?'
'Yes, absolutely.'
'Cannot you skip some part?'
'No, it is impossible.'

But then, Shelley was defined by his own histrionics. When the nineteen-year-old poet had gone to Ireland, determined to stir radical opinion, his father-in-law-to-be, William Godwin, warned him, 'Shelley, you are preparing a scene of blood!' And when Robert Southey accused him of having driven his first wife to suicide, the poet, in provocation as much as defence, retorted, 'I could tell you a history that would open your eyes.' As Stephen Hebron, a contemporary Shelleyan, tells me, the modern world would surely have diagnosed mental illness in his intense mood swings, heightened by opium. Shelley was now taking laudanum regularly, showing the bottle to Thomas Love Peacock and saying,

'I never part from this.' Peacock in turn parodied him as Scythrop Glowry, 'troubled with the *passion for reforming the world*', and living in Nightmare Abbey, over a shore dashed by monotonous waves; a site 'ruinous and full of owls' and staffed by a butler named Raven and a steward called Crow. And if Shelley's psychodrama read like opiated Shakespeare, then he identified deeply with *The Tempest*, his favourite play. Its influence runs powerfully through his most ambitious work, *Prometheus Unbound*.

Shelley's sprawling epic, completed in 1816, opens with Prometheus, punished by the gods for having given fire to humanity, chained to 'a Ravine of Icy Rocks in the Indian Caucasus', a setting that echoes the Arctic scenes of *Frankenstein* (itself 'A Modern Prometheus'). Bound to 'this wall of eagle-baffling mountain', he faces a world locked in a thousand-year winter – 'The crawling glaciers pierce me with the spears | Of their moon-freezing crystals' – while 'Heaven's winged hound', a hawk, pecks his innards with a poisonous beak. This is a nightmare Narnia, a very bad trip indeed.

The play swells with spirits, phantasms, furies and fauns who speak ritually, like Prospero and his familiars. Both Prospero and Prometheus are captives; Shelley describes a kind of natural Promethean fire in marsh gas or will o' the wisp: fire out of water. The Earth and Ocean are listed as *dramatis personae*; the elemental setting is apocalyptic, closer to Turner's mythic miasmas or John Martin's lurid panoramas, a science-fiction scenario of 'those million worlds which burn and roll | Around us', of dark seas 'lifted by strange tempest' and places peopled by 'Terrible, strange, sublime and beauteous shapes' whose 'bright locks | Stream like a comet's flashing hair' as they sweep onward. Such cosmology could prophesy the beginning of all matter, or the ending to 2001. The magus Zoroaster meets 'his own image walking in the garden', and instead of devils, 'all the gods | Are there, and the powers of nameless worlds'. Veering between east and west, past and future, particles and universes, a future magician, J. Robert Oppenheimer, might step in to declare, 'Now I am become Death, the destroyer of worlds.'

Here too is the crouching incubus of Fuseli's *Nightmare* – 'So thy worn form pursues me night and day' – lying in wait like Tan-yr-allt's demon; little wonder that Edgar Allan Poe so admired Shelley's play. Its characters stand shadowed against the sublimity of the physical world; their own creator did not quite comprehend their meaning, as if he were possessed by them, as much as he had made them. They cross time zones and technologies: Asia, the Oceanid beloved by Prometheus, sails in an all-seeing airship 'over a sea profound'.

Shelley was writing at a time when much of the planet remained unmapped. The earth still had room for such imaginings, as if there existed, in some sealed library, a metaphysical globe in which reality and poetry could merge. As the synapses snapped inside his head, his pathology zoomed from the universal to the microcosmic; a hugeness collapsing on itself, blood pulsing and flowing like the tides. Out of a world of his own making, he addresses what we mean in relation to Nature; our place within it, and its place within us. He seems suited to neither. Instead, he moves in between.

Surrounded by real and imaginary terrors, the poet's fate was foretold by water, written in it, like Keats's name. Two years after their marriage, Shelley left his first wife, Harriet, with whom he had eloped when she was sixteen, for Mary Godwin, who was also sixteen. Three years later Harriet, who had so regretted the loss of the fine, bold sea at Portmadoc, threw herself into the brown-green Serpentine. She drowned in the goose-shit-stained waters where I swim when exiled to the capital for the day, pushing through the murky lake where Virginia Woolf recorded the drowning of another young woman while imagining herself plunging into its mud. When Harriet's body was found, a month after she had gone missing, it was discovered that she was 'far advanced in pregnancy', the result of an unwise union with another lover.

Only two months before, Fanny Godwin, Mary's stepsister, had taken her own life too, with an overdose of opium; their mother, Mary Wollstonecraft, had also tried to drown herself in the Thames in 1795, throwing herself off Putney Bridge when she discovered that her lover had left her. All these falling bodies stirred bitter remorse. 'What storms then shook the ocean of my sleep,' Shelley wrote, 'And how my soul was as a lampless sea | And who was then its Tempest.'

Fleeing England in 1816, Shelley and his circle convened at the Villa Diodati on Lake Geneva, in the shadow of the Alps and their slow-moving rivers of ice rolling down from the mountains where Prometheus might be bound. In a year without a summer – thanks to a volcanic eruption thousands of miles away which darkened the skies by day – they invented their myths for an industrial age already occluding the heavens on its own account. Out of that lakeside another primordial Creature was born, as if brought to life by lightning striking the water; a beautiful man become a Caliban.

A few days later, Shelley and Byron were out on the lake when their boat was nearly swamped in a storm. Shelley refused to abandon the vessel, despite Byron's pleas; unlike his fellow poet, who was celebrated for his aquatic prowess, Shelley could not swim at all, and instead held fast to the boat. Naturally, he rationalised the incident in irrational terms: 'I felt in this near prospect of death a mixture of sensations, among which terror entered, though but subordinately. My feelings would have been less painful had I been alone; but I knew that my companion would have attempted to save me, and I was overcome with humiliation, when I thought that his life might have been risked to preserve mine.'

As Richard Holmes, his biographer, noted, 'the most extraordinary thing is that after this incident, Shelley did not ask Byron to teach him to swim'. It was a fatal neglect. 'If you can't swim, | Beware of Providence,' as Shelley himself wrote. But then, his own history did not appear to teach Shelley anything.

Shelley sat on the banks of the river, watching Trelawny performing a series of aquatic manoeuvres he claimed to have learned from South Sea islanders. The poet was astonished, and envious.

Edward John Trelawny was a colourful character, even among Shelley's cast of theatrical personalities. Born of Cornish stock as his name suggested, Trelawny had a fiery temper and a piratical manner. As a young boy he had killed, torturously, his father's clipped-winged pet raven, beating it until its eye hung out of its socket; when he was sent to boarding school, he was expelled for trying to burn it down in revenge for the floggings he had received. He joined the navy in 1805, serving on the fighting *Temeraire* – bitterly regretting that he had missed the Battle of Trafalgar – and sailed as far as the East Indies, while remaining dismissive of discipline: on one occasion he assaulted his superior officers. Now at a loose end in Italy, Trelawny had befriended Byron at the height of his infamy, this seducer of whom Lady Caroline Lamb said, 'this pale handsome face holds my destiny'. They suited each other, these two adventurers: Trelawny became known as Lord Byron's Jackal, and together the two men boxed, fenced and swam.

It was through Byron that Trelawny had met Shelley, and fell in love with Mary's half-sister Claire Clairmont (who'd already thrown herself at Byron in the most wilful manner). Trelawny was well placed to observe the strange relationship between Shelley and Byron: one willowy and neurotic, the other determinedly physical but equally conflicted; both to be made legendary, not least by Trelawny's stories.

Equally, you can see why the poets were drawn to Trelawny. He represented action, rather than introspection. Writers wrote 'to console themselves for not living', said André Maurois; Mary Shelley expressed it even more vividly: 'Trelawny lives with the living, and we live with the dead.' Swaggering onto their scene, suffused with his own mystique, the former naval officer was powerfully built, six foot tall with dark piercing eyes, a handlebar moustache and a tanned face which reminded Mary of a Moor. She called him 'a half Arab Englishman', with an extravagance 'partly natural ... and partly perhaps put on', forever telling 'strange stories of himself, horrific ones'. We might find his figure to be almost camp, the stuff of mustachio'd melodrama, a coloured cut-out waved about in a cardboard theatre. But there was also another charge at work among these identity-challenged new romantics. Later critics would claim that Trelawny was in love with Shelley, and perhaps Byron too, and that Mary was a lesbian. Trelawny himself said that Thomas Jefferson Hogg, with whom Shelley had lived in London, was the poet's 'one true love'.

Whoever owned to these desires, southern Europe, its cities, ports and shores, offered licence enough to allow transgression and performance (Byron saw Venice as a 'sea-Sodom', filled with 'marine melancholy'). In that Italian summer of 1822, Trelawny made it his business to get to know Shelley and Byron. His observations of the two men are among the most vivid of all contemporary accounts, although they blur into his own mythomania. Like Shelley, George Gordon, Lord Byron, had the air of a royal exile, a prince driven out of his homeland – not by politics but by rumours of incest and 'unnatural crime'; Caroline Lamb cited Robert Rushton, the handsome young man who accompanied the lord as his page, as one of those whom Byron had 'perverted'. With a dog-like loyalty, Rushton slept in a cubbyhole next to his master at Newstead Abbey; a famous image of the pair, poet and page, shows them ready to set out to sea – and hints at something more.

The poet's friend John Cam Hobhouse teased him for such recklessness, telling Byron that he might get shot for advertising his criminal relationship so boldly. Before leaving England for Lisbon in 1809, Byron stayed in Falmouth, from where he wrote of his attachment to Rushton – 'I like him, because, like myself, he seems a friendless animal' – and boasted of his seduction of another boy in the town, scratching his own name into the window of Wynn's Hotel as if to mark his conquests. In his private correspondence as in his public image, Byron made it clear that he didn't care what anyone else thought. An aristocrat operated under privilege – private law – and a poet was further set outside the normal world. He was as entitled to have his affections immortalised as he was to commission a portrait of an animal he loved even more.

Byron's nautically-named dog, Boatswain (the word comes from the Old Norse, *sweyne*, a young man in charge of a vessel), was bred to the water – Newfoundlands have webbed feet, like their Labrador cousins – and became Byron's companion in Newstead's wide and weedy lake, where he sometimes assumed his master was drowning and tried to pull him out. Boatswain's loyalty exceeded that of any human bosun; he looked quite as pugnacious and aristocratic as his owner. After his death in 1808 from rabies – ironically known as hydrophobia – during which Byron nursed

him, disdaining any fear of infecting himself (Virginia Woolf, ever attuned to animals and their owners, suggested that 'Byron's dog went mad in sympathy with Byron'), Boatswain was buried at Newstead and honoured with an elegy from his master to his 'firmest friend | The first to welcome, foremost to defend'.

Many, if not most of Byron's retainers were furred or feathered. He walked with them like a god. When he left England in 1816, lines of onlookers stood either side of the gangway at Dover, watching as his sofa, books, china and glass were decanted onto his ship. Roaming Europe in his enormous coach, copied from one used by Napoleon, the poet was accompanied by a menagerie. In Ravenna, visitors to Byron's rented villa were greeted by eight enormous dogs, three monkeys, five cats, an eagle, a parrot and a falcon, all of whom fought with one another fitfully, 'and made up as it suited them'. When he moved to Pisa, his procession consisted of five carriages, six manservants, nine horses, and an assortment of dogs, monkeys, peacocks and ibises. It was as if he were lord of all the wilderness.

No one could ignore the passing of this English aristocrat, as resplendent as any exotic bird or beast and surrounded by his familiars, sweeping all before him and them. Yet in Trelawny's account Byron, then in his mid-thirties, is a surprisingly vulnerable figure, for all his glamour. At five foot eight and a half, he is not tall, and he shuffles as he walks, because of his club foot. It was his equivalent of Shelley's high voice, an Achilles heel which made the rest of the show seem more extreme. He dressed majestically in a braided tartan jacket, blue velvet cap with gold braid, and 'very loose nankeen trousers, strapped down so as to cover his feet'. His vanity was apparent, and included the use of curl papers for his hair and cosmetics for his face, and diets to maintain, in a modern manner, his body image. Declining Shelley's obsession with 'mystifying metaphysics', his aura was one of action and even mayhem (even Shelley admitted that his friend was 'mad as a hatter'). Turning his bodily self into a paradox, Byron preferred to refer to his disability as 'cloven footed, diabolic rather than infirm' – evoking the demon many believed him to be – and when he checked into a hotel, he wrote in the register against his age, 'One hundred.'

'The Devil,' Byron declared, as if watching himself limp onto the stage, 'is a Royal Personage.' He dallied with the same Satan whom Shelley feared; and where Shelley saw himself as a 'Lost Angel of a ruin'd Paradise', Byron assumed the persona of the stalking, or perhaps hobbling, fiend. Most of all, he sought to transcend his terrestrial lameness. 'In the water a fin is better than a foot,' Trelawny affirmed, as if the lord were halfway to becoming a porpoise; 'and in that element he did well; he was built for floating – with a flexible body, open chest, broad beam, and round limbs ... If the sea was smooth and warm, he would stay in it for hours.'

Like his dog, Byron had been bred to the water. His family crest was surmounted by a mermaid, and his grandfather, Vice Admiral John Byron, was known as Foul-weather Jack on account of his reputation for attracting storms. Byron himself was said to have been born with a caul, which doubtless emboldened him in his feats. At college he had swum in the Cam, often with his friend Skinner Matthews, who later drowned there, having become entangled in the weeds. At twenty-one, Byron swam the Thames from Lambeth to Blackfriars. But if the water was his liberation, it might also have been a cover for his true nature, just as his strenuous assertions of masculinity – his boxing, his adventuring, and perhaps even his womanising – were compensations or disguises, another kind of veil.

Having left England under the cloud of a scandal that could not be named, Byron swam through Europe. In Lisbon he swam the Tagus. In Athens he swam daily at Piraeus and saved a girl, sewn up in a sack as a punishment for adultery, from being thrown into the sea. In Venice he swam the length of the Grand Canal. And in his greatest feat, he swam from one continent to another, across the Hellespont, although the young marine who accompanied him later drowned too. Byron exhorted, 'Roll on, thou deep and dark blue ocean – roll!' The sea allowed him to challenge his mortal state; it made him alive and free. 'The great object of life is sensation,' he said, in words that augured Wilde's, 'to feel that we exist, even though in pain.'

One day Byron dared Trelawny to join him on a swim from the shore back to the poet's yacht, Bolivar, and to dine in the sea alongside her.

Trelawny, a better swimmer, reached the boat long before Byron, and ordered the meal to be served on a grating. Treading water all the while, the two men ate their buoyant dinner, washed down with a bottle of ale and followed, for Trelawny, by a cigar, to which Byron objected violently. Then they set off for the shore, but a hundred yards out the food and drink backfired on Byron, who began to retch and to suffer severe cramp.

Trelawny told his friend to place his hand on his shoulder so that he could help him back to safety.

'Keep off, you villain, don't touch me,' Byron replied. 'I'll drown 'ere I give in.'

'A fig for drowning,' Trelawny said; 'drown cats and blind puppies.'

'Come on,' Byron shouted. 'I am always better after vomiting.'

They returned to the boat, but Byron sat on the ladder, refusing to get on board, demanding that they should swim back to the shore. Trelawny said he'd had enough.

'You may do as you like,' Byron replied, and jumped in the water. He was ill for two days afterwards, and had to take to his bed.

In these escapades, Trelawny portrayed himself as a solo audience for the poets' performances. Back on the banks of Arno, where we left him with Shelley, Trelawny completed his Indonesian exercises, climbed out of the river and began to dress himself on the shore.

'Why can't I swim, it seems so very easy?' Shelley asked.

'Because you think you can't,' Trelawny replied, patiently. 'If you determine, you will,' he continued. 'Take a header off this bank, and when you rise turn on your back, you will float like a duck; but you must reverse the arch in your spine, for it's now bent the wrong way.'

Any onlooker could have predicted that this lesson would not end well. Shelley pulled off his clothes and jumped in the water, but instead of following Trelawny's instructions, he dropped to the bottom and lay there like a languid eel, making no attempt to struggle or save himself. He had sunk deliberately, as if seeking his amniotic origins, an unplugged foetus. Had he been born with a caul like Byron, he might have been protected; but Shelley wanted to shed his earthly state entirely.

Trelawny – by now used to these mad antics – dived in to save the poet who, he claimed, would have drowned if he hadn't been fished out. Recovering back on land, Shelley said, 'I always find the bottom of the well, and they say Truth lies there. In another minute I should have found it, and you would have found an empty shell. It is an easy way of getting rid of the body.'

'It's a great temptation,' he added; 'in another minute I might have been in another planet.'

Some thought he might have come from one. Shelley seemed a 'true alien', in the words of a modern critic, Karen Swann, a stranger or visitor 'who speaks with no natural voice and is animated by no natural life'. Nor did any of his near-disasters discourage him. In Italy he nearly drowned in a canal, yet still bathed in ponds 'as transparent as the air' and 'exceedingly cold', as he boasted to Peacock. He would undress and sit on rocks over the pool, reading Herodotus 'until the perspiration has subsided', then leap into the water, climbing up the waterfalls and 'receiving the spray all over my body'.

On one occasion Trelawny went searching for him in a wood, and was taken by an old man to a trail of books and papers and a hat, and beyond that 'a deep pool of dark glimmering water'. The old man said, '*Eccolo!*', which Trelawny took to mean that Shelley was in or under the water.

A strong light shone through the pines, one of which had collapsed into the pool. The wind rushed through the trees. When

Trelawny finally found Shelley, busy writing verses while strumming his guitar, the poet asked him, 'Don't you hear the mournful murmurings of the sea?' It was a scene which could have come from one of Ovid's Tales or a film by Michael Powell and Emeric Pressburger. Picking up a fragment of the manuscript which lay scrawled and blotted on the ground, Trelawny could read only the first two lines: 'Ariel, to Miranda take | This slave of music.'

Shelley was determined to spend that summer by the sea. He and Mary, and their friends Edward Williams – like Trelawny, a former naval man – and his wife Jane, had come to the Bay of Spezia, the Bay of Hope. No suitable palazzo could be found for Byron and his entourage there, but after some searching the Shelleys and the Williamses discovered a house for themselves at Lerici.

Casa Magni, a former monastery, could have stood no nearer the shore. It was built over the sea wall, its ground floor left unmade because it flooded at high tide. The house was half sea itself, like the semi-submerged apartment constructed by the renegade nineteen-sixties scientist John C. Lilly to prove how like dolphins we were – or vice versa – by allowing his researchers to live with their subjects twenty-four hours a day. 'We all feel as if we were on board ship,' Captain Williams wrote, '– and the roaring of the sea brought this idea even into our beds.'

At first Shelley loved it; Trelawny saw him as a South Sea islander, delighting to sport in the water. The poet said, 'I wish I was far away on some lone island, with no other inhabitant than seals, sea-birds, and water-rats,' and told Trelawny that instead of wasting his life reading Greek and Latin, he ought to have learned swimming and sailoring instead. To that end, he resolved to acquire his own boat. But when Captain Roberts, who was to build it, came to lunch, he was entertained by the three women of the house; Shelley was nowhere to be seen.

Suddenly, the poet appeared in the shadows at the back of the room, trying to creep upstairs. Challenged by his wife on his behaviour, Shelley approached the table to explain. He didn't have to: he was naked and fresh from the sea, a selkie who'd shed his sealish skin. The women affected to avert their eyes. 'Yet he was good to look at,' André Maurois imagined, 'his hair full of seaweed, his slender body wet and scented with the salt of the sea.' But Shelley's dreams were increasingly filled with images of disaster. They came in like the tide. Edward Williams was standing on the terrace with Shelley when the poet grabbed him by the arm and stared down at the waves, saying, 'There it is again! – there!' He'd seen a naked child rising out of the sea, clapping its hands and smiling at him.

'Shelley *sees spirits* and alarms the whole house,' said Mary. Like Tan-yr-allt and the Villa Diodati, Casa Magni seemed to raise ghosts from the water; the former monastery had become Nightmare Abbey. In another terror, Shelley saw Edward and Jane 'in the most horrible condition, their bodies lacerated – their bones starting through their skin, the faces pale yet stained with blood'. In his dream, Edward told him, 'Get up, Shelley, the sea is flooding the house and it is all coming down,' and as the poet rose and went to his window, he saw the sea rushing in and himself strangling his own wife. The water he had tried to control in Wales was now lapping at his door, and Shelley bore witness to the most awful apparition yet: a cloaked figure that came to his bedside and which he followed to the hall, where it lifted its hood to reveal his own face, asking, '*Siete satisfatto?*' – Are you content?

The fear was infectious. Jane Williams claimed that while Shelley was out sailing she had seen him walking on the terrace, twice. This hyped-up atmosphere of signs and wonders was stirred by a sequence of real tragedy: Shelley's young daughter Allegra had just died of typhus; Mary was about to miscarry; Byron's illegitimate daughter by Claire died. These Shelleyan orphans seemed to announce their fatherlessness even before they were born. There was no consolation for any of them in this world, least of all in the sea. It did not care. Nothing could help them. Everything was speeding up; life and death were collapsing in on each other.

On the afternoon of 8 July, Shelley set sail across the bay on a twenty-four-foot, twin-sailed yacht, one of a pair designed by Trelawny and built by Roberts for him and Byron. Shelley had intended to call his boat *Ariel*, but when it arrived, Byron had already christened it *Don Juan*, and the name was painted in garish letters on the forward mainsail. Shelley insisted it should be removed; he didn't want to sail under his friend's most infamous creation. The fact that Byron's poem included the sinking of the ship which carried his immoral hero was hardly a good omen, either.

Shelley, three weeks from his thirtieth birthday, was dressed in maritime fashion, in a short double-breasted reefer jacket, white nankeen trousers and black leather boots. He was accompanied by Edward Williams and an eighteen-year-old English boy, Charles Vivian, 'a smart sailor lad' who'd come with the yacht when it arrived from the shipyard; he was 'quick and handy, and used to boats'. That afternoon, they were heading back home from Leghorn harbour. Trelawny had set sail alongside them on Byron's *Bolivar*, but soon lost sight of *Don Juan* in a thick sea fog. One of his crew looked at the gathering clouds and said, 'The devil is brewing mischief.'

By six o'clock the storm had struck. A passing ship told them to reef in their sails or they would be lost; Shelley was heard to shout in his shrill voice, 'No,' and prevented Williams from doing so.

The boat sank under full sail. Later, there would be claims that it had been rammed by raiders who thought it was laden with Byron's gold. It was five days before the wreck was discovered, and ten before the bodies were found, three miles apart. 'The sea, by its restless moaning,' Mary recalled of those suspended days in which Shelley neither existed nor did not exist, 'seemed to desire to inform us of what we would not learn.' She waited like her biblical namesake at the tomb. The atheist poet had wanted to sleep at the bottom of the sea. The sea would not let him.

Williams was discovered undressed, apparently in the act of attempting to swim, although the sea habitually strips its victims. Since he could not swim, it was assumed that Shelley had drowned at once; he'd told Trelawny that if their ship did wreck he would sink immediately, to save others the danger of trying to rescue him. Almost unrecognisable now, he was identified by his nankeen trousers and white silk stockings, and a volume of Keats's poems – the book was found stuffed, as if hurriedly, in his pocket. The third of the party, young Charles Vivian, was found on a far beach. The official report stated that his body was clad in a cambric shirt, cotton waistcoat, blue-and-white-striped trousers; the image of a neat sailor boy made horrific by the fact that the head had been eaten by fishes. Not for Vivian the rites which would honour his renowned, drowned shipmates. His loss was barely noticed, in the corner of the picture. He was buried in the sand, like a clam.

Williams's and Shelley's remains were also temporarily interred on the beach, ready to be cremated there; partly because Italian law demanded it in order to guard against plague, and partly because the bodies were so badly mangled. Trelawny, with a mixture of the practical and the piratical, devised the pagan ritual by which they were to be disposed. He had a furnace made of iron bars and sheet iron, a kind of human oven. Trelawny, Byron and Leigh Hunt assembled on the strand. The spectacle had drawn a crowd, including a number of richly dressed ladies. Soldiers gathered the fuel: brushwood and driftwood, flotsam from former wrecks. Frankincense, salt, wine and oil were thrown on the pyre, as if some classical hero were being sent off to the underworld at the outer bounds of the ocean.

As the flames consumed Williams's body – its meat more like the carcase of an animal than that of a man – Byron announced, 'Let us try the strength of these waters that drowned our friends. How far out do you think they were when their boat sank?' At that, he stripped off and waded into the sea, followed by Trelawny. They swam nearly a mile out before Byron, as usual, got sick, and they had to return to shore.

Shelley's assumption at Viareggio the following day was even more dramatic. Trelawny claimed that 'The lonely and grand scenery that surrounded us so exactly harmonized with Shelley's genius that I could imagine his spirit soaring over us,' while Leigh Hunt was reminded of an airy spirit, 'found dead in a solitary corner of the earth, its wings stiffened, its warm heart cold'.

They were about to burn Icarus. In attendance were the same group of grieving friends commemorated years later in Louis Édouard Fournier's famous painting – a fanciful, preternatural scene, since it included Mary Shelley kneeling on the shore despite the fact that she was miles away that day.

As the poet's remains were dug up, Trelawny felt they were no more than wolves or wild dogs scavenging Shelley's corpse, 'tearing out his battered and naked body from the pure yellow sand that lay so lightly over it, to drag him back to the light of day'. At one point the shovel struck Shelley's skull with a dull hollow sound. His corpse was now 'of a dark and ghastly indigo colour'. Trelawny had brought his 'iron machine', and the same rites were observed; he noted that more wine was poured over the poet's dead body than his living body had ever consumed. This, and the oil and salt, turned the flames yellow as the air was made wavy and tremulous by the Promethean fire. Shelley was a fallen comet, still smouldering on the beach. Then the corpse fell apart, and the brains bubbled in their cranial cauldron.

Byron, unable to bear the sight, had already taken off all his clothes again to swim back to his yacht. Meanwhile, back on the gothic barbecue, Shelley's heart resisted the flames. Trelawny snatched it from the fire, and when the embers had died down, swept the poet's ashes into a box and returned to *Bolivar* himself.

Were these real people at all? Once Shelley had drowned, the *dramatis personae* were dispersed, as if they had only been held together by his magic. Trelawny proposed marriage to Mary Shelley, but she declined, saying that her name would look pretty on her tombstone. Hogg proposed to Jane Williams – who had to reveal that she had never been married to Edward Williams, and still had a husband somewhere in India.

Byron became ever more Byronic, pursuing his championship of the Greeks and funding his own private army to free them from the Ottoman Empire. Trelawny accompanied him partway on the journey, giving up Benjamin, his African-American groom, to Byron – his blackness 'a mark of dignity' – as well as donating his green embroidered military jacket to the lord. Also at Byron's side was his other loyal accomplice – Lyon the Newfoundland, who had replaced Boatswain. Throughout the voyage, Byron and Trelawny swam every day, diving off *Bolivar* at noon, 'in defiance of sharks or weather', accompanied by Lyon, who like his predecessor was ever ready to leap into the water to join his master. It was the only exercise Byron got, since his disability made it difficult to walk on deck, stumping about like Ahab.

Sailing into battle in a new heroic outfit – crowned with a shiny Greek helmet with a great feathery crest like a bizarre bird – Byron would meet his end, which he had long predicted, not drowned as his friends had been – since he had been born veiled – but expiring in a dreary villa next to a shallow, slimy salt marsh, where he wrote his final poem, an appeal to the last of his boy loves, fifteen-year-old Loukas Chalandritsanos: 'Such is this maddening fascination grown, | So strong thy magic or so weak am I.' The pain of this unrequited affection fatally lowered the lord's spirits. Less than two years after Shelley's drowning, Byron, who insisted on riding in the rain, caught a chill. He died of a fever on 19 April 1824, during a violent electrical storm, as if Ariel had come for him. He was thirty-six years old.

To close the circle, Trelawny composed another macabre scenario which, like almost everything he ever said, may or may not have been true. Wading through the water, he arrived at Byron's pathetic lying in state five days after his death. His friend looked more beautiful than he had in life: the lines on his face had fallen away, the skin drawn back; no marble bust could match its pallor, his fate. And yet Trelawny remembered how dissatisfied Byron had been with his physical self, and how he had longed to cast it off, too. Sending the lord's faithful servant, William Fletcher, for a glass of water, Trelawny, left alone in the room, pulled back the sheet that covered the corpse, and discovered the true extent of Byron's deformity.

Even though, by his own account, Trelawny must have seen his friend naked many times – once, after swimming, Byron held out his right leg and said, 'I hope this accursed limb will be knocked off in the war,' to which Trelawny replied, 'It won't improve your swimming' – the poet's body was shockingly revealed in this last scene. Not only was his right foot clubbed, but both legs were mis-shapen and withered to the knees. The dead Byron had the 'form and features of Apollo', but 'the feet and legs of a sylvan satyr', half hero, half animal.

Trelawny reeled back, retrospectively working out this con-juring trick, this last transformation. He'd always thought Byron exaggerated his disability for effect, boasting of a body 'scarce half made up'. Now he recalled, as if in a flashback, the wiles Byron employed to disguise his true self: high-heeled boots with uneven soles, their toes stuffed with cotton wool; trousers tailored wide below the knee and strapped under his instep to cover his twisted legs and feet; the way Byron would half-run, half-totter into a room. His incompleteness was somehow shameful; it seemed to unman him. His constant battle with his weight was part of the same process; Trelawny noted that much of the time that he knew him, Byron existed in a state of semi-starvation. If he grew too heavy, the pressure on his deformed legs meant he could barely walk. What a fragile artifice he seemed, this bone-weakened body, this bastard self-construction, this charming man who seduced women but loved boys. Little wonder he'd left England to seek the solace of the sea. For Byron, so ill-suited to the land, the water was the only place where he could be himself.

After many more adventures, violent and amorous, surviving an as-sassination attempt and siring a daughter by a child bride, Trelawny retired to Putney, espousing vegetarianism and temperance and cold water, eschewing underwear and overcoats and hot food. Yet his wanderlust drew him to new exploits: a vague attempt to found a utopian commune in Virginia, and a stupendous, if not stupid, swim across the Hudson above Niagara Falls, nearly dying in the process. 'Seeing the land on each side I thought it absurd to be drowned in a river. I heard the voices of the dead calling to me. I actually thought, as my mind grew darker, that they were tugging at my feet.' It would have been, he thought, 'a fitting end to my wild meteor-like life'.

Even in his final retreat, to the genteel south-coast village of Sompting in Sussex – where he swam a mile in the English Channel every day, whatever the weather – Trelawny played out his role. He padded about barefoot, chopped his own wood, drew water from his well. He also received young admirers such as the sandal-wearing Edward Carpenter, who shared his vegetarianism and disdain for 'bathing-drawers' (and had recently returned from Massachusetts, where he'd paid tribute to Thoreau by swimming in Walden Pond). Trelawny remained a republican and an atheist, yet like Byron and Shelley he was assumed into Victorian legend, partly through his own mythomania, partly through art. In *The North-West Passage*, painted in 1874 by John Everett Millais and subtitled 'It might be done and England should do it', the grizzled and bearded sea-king is seen seated beneath a portrait of Nelson, hero of a battle Trelawny had never seen. There are draped ensigns and marine charts, and a glass of grog to hand (the abstemious subject protested at this detail, and threatened to challenge Millais

to a duel). A young girl, supposed to be Trelawny's daughter, reads at his feet. The reference to The Tempest seems clear. And as we look over his shoulder, through a window and out to sea, we see a single yacht, passing by in the distance, with somewhere to go.

In the fast-running waters of the Tavy which rush off Dartmoor, accelerated by its granite and turned brown by its peat, I swim with Tangle the retriever, both our heads held high. He paddles with webbed feet, and occasionally dips his greying muzzle down into the water like a bear, nosing about the riverbed. Someone leans over the bridge and shouts, 'Very Lord Byron.' Tangle climbs out and shakes his coat, from sleek head to shaggy tail.

Tangle is my Boatswain. He puts me in mind of a story told by Caitlin Davies, of a race in 1880 in the Thames between one R. Smith, 'a known aquatic performer', and a six-year-old black retriever named Now Then, who had previously rescued seven people from drowning. The dog won easily, still swimming towards London Bridge after two hours, long after his human rival had fallen far behind. An etching shows the dog's profile held above the waves, just as Byron's dogs swam heroically with him.

Tangle is my braveheart, running up to the tor, his black coat flying like a Renaissance prince, with a twisted green silken rope for his leash. He is often approached by other dogs, yet is quite unassuming in his handsomeness, barely acknowledging, with a sideways glance, the admiring stares of passersby as he walks on. They are envious of my elegant escort, my charismatic canine companion. He cannot be far from water for long. He runs back to the river, pausing to ensure that he knows where I am, and vice versa. This is his domain. Most noble of his breed, he dashes over moors where his wolfish ancestors prowled. Down in the valley of the Dart, where maple leaves drift past us like yellow stars in the dark river, Tangle dips his snout, seeing what I do not, raising his head to shake and snort, clearing his nostrils for a second try. He sits beside me in the autumn sun, sniffing the air, scenting worlds beyond my senses, tolerant of my ineptitude, ignorant as I am of the reality around me. He knows the hour of my coming and my going.

Dogs of course live faster, more intense lives than ours, cramming in every experience, rendering ours dull and sluggish in comparison. They experience ecstasy and despair every minute while we waste time, suffering over tediously elongated spans, hanging around long after our expiry dates. We may think them bound to our will, but dogs know we are bound to theirs. 'For the dumb creatures,' as Orlando, keeper of her own elk-hounds, was aware, 'are far better judges both of identity and character than we are.' I wonder what Tangle remembers; or if he is shackled by memory at all, in the way we humans are cursed with the knowledge of our mortality.

Mary Shelley 'remained trapped by memories both idealized and remorseful', as Richard Holmes would write, although her life after Shelley 'attained a curious stillness, interrupted only by sea-bathing at Sandgate', in Kent. Her husband's ashes had long since been interred in Italian earth, in the same churchyard in Rome where John Keats lay. Keats had asked for one line on his stone, 'Here lies one whose name was writ in water.' On Shelley's, Trelawny had Ariel's song inscribed.

> Nothing of him that doth fade,
> But doth suffer a sea-change
> Into something rich and strange.

Years later, Trelawny's own ashes would be laid loyally by Shelley's side.

All that clamour has long since fallen quiet, like the lull of a tide, leaving only memorials standing proud. In Oxford, Shelley's half life was commemorated – by the college that expelled him – with a weighty sculpture by Edward Onslow Ford.

The naked poet – for whom Ford's own fourteen-year-old son Wolfram had modelled – lies splayed over a sea-green slab, supported by bronze-winged lions and mourned by a female muse. Shelley is rendered in marmoreal white, languidly draped as much in an opium daze as in drowned demise, his frail body carved out of ten tons of crystalline limestone, itself formed from crushed sea creatures.

Despite its intersexual air and echoes of the poet's preoccupation with the classical hermaphrodite, nothing could be less Shelleyan than this unwieldy Victorian lump, donated by his daughter-in-law, Jane, Lady Shelley, to University College in 1893. It is enclosed under a blue dome scattered with gold stars – just as the somewhat exuberant Lady Shelley maintained a violet-scented Shelley Sanctum in her Bournemouth house in which, lit by a red sanctuary lamp, were preserved locks of his hair and blue pots filled with slivers of his bone. Another jar was said to contain part of his heart, wrapped in a page torn by Mary from her copy of his *Adonaïs*. In her devotion, Jane had made the pilgrimage to Casa Magni hoping to experience a vision, and claimed an Italian peasant told her of Shelley, 'he was like Jesus Christ. I carried him in my arms through the water – yes, he was like Jesus Christ.' She was clearly thinking of the memorial to Shelley in Christchurch Priory, where he lay in Mary's arms as another pietà, a trail of seaweed wrapped around his arm.

Oxford's version lies locked behind an iron grille, to which Lady Shelley was given a golden key. On its unveiling, the memorial was enthusiastically received by one commentator: 'It is as if the restless sea, in whose breast he had been tossed all those years, had laid him at last upon the threshold from which he had been first cast forth.' A more sceptical critic was less impressed – he described the sculpture as 'a slice of turbot laid out on a fisherman's scale'.

Having made my way through the city's intimidating stone lanes to this cloistered shrine, I'm surprised at its scale. It's much larger than I'd expected: the size of a hall in a stately home, its gates and bars imprisoning the poet who was gated and barred from this place. The memorial is set a few steps down, below ground level, forcing the visitor to descend as to a saint's effigy in a crypt. Once the chamber was flooded by undergraduates as a prank, turned into a pond, complete with goldfish. I'd like to have seen this space glinting with fish like a grotto, the light of the waves rippling on its sky-dome; a miniature Casa Magni, lacking only a few sporting dolphins from the Tethys Sea. But this is a wrecked casualty, wheeled in on a trolley. In a corner of the chamber there's a radiator fixed to the wall, as if to warm up the corpse.

This is no place for Shelley, drowned or dry. 'The dead were no longer able to object or to disgrace themselves,' as Denton Welch wrote. 'Once the power to ridicule or to degrade them had been taken away from them they could be manipulated; they were ready then for honour and remembrance.' As I peer through the cage, the sprawled Shelley becomes a broken bird, tossed in the stormy surf and thrown onto the shore.

I cross the busy street and descend into the gloomy basement of a nearby building while the city's cyclists and pedestrians weave above me. The bunker contains a conference table, an overhead projector, and a whiteboard on which some explanatory notes have been lately wiped away. Anything from health and safety to philosophy to procedures for the evacuation of priceless treasures might be discussed down here.

I wait for Stephen Hebron to return from a distant store. He arrives carrying a small grey cardboard box. Its catalogue-cum-convict's number – MS Shelley adds. e. 20, fol. 721 – gives few clues to its contents. The package contains a package within a package within a package; Stephen unwinds the cord that ties the first, unfolds the flaps that envelop the second, then unsheathes the sea-green leather and gilt-blocked slipcase which reverentially protects the last. The ceremony is somewhat undermined by the final object: a block of what looks like blotting paper, loosely stitched – or rather, coming apart – at its exposed spine.

Laid on a pair of anodyne grey foam pads, this precious arte-fact might as well be sopping wet, since it is all that remains of the notebook which Shelley took with him on his last voyage. Fished from the wreck when the boat was raised, it comprises fewer than fifty pages of Shelley's workings, doodles, sums and summations in pencil and ink. Amazingly, and for all the control extended around this object, I am permitted to handle it without gloves. It is like having unprotected sex with a book.

Soaked, splattered, stuck together; washed-out, faded, mil-dewed and bleached by its improbable survival since leaving the poet's hand and being subjected to history's spin-cycle, these pages are more pattern than writing, more kelp than literature. Like the sodden sheets Trelawny found around the poet by his Italian pool, they turn as rags in my hands, their fine laid paper made of pulp and cloth, layered with meaning and unmeaning. The brown ink bleeds into itself or disappears entirely mid-word, erased by some censoring agent, as if the authorities had finally redacted the rebel's mad treason.

Now in climate-controlled captivity, it will never see another storm. Its crumpled, frayed, salted and sanded leaves have been flattened out and conserved, catalogued and boxed up – for our safety, as much as its own, like nuclear waste. Pages which endured the mighty tempest are too fragile to see the twenty-first-century light of day. Too late, too soon, too powerful, this collection of incantations. 'And, deeper than did ever plummet sound,' Prospero says, 'I'll drown my book.'

Shelley's italic hand slopes into itself, colliding one letter with another, trying to catch up with his ideas. The book – but it is no mere book, one of those bound prisons of words – is filled with his elegy to Keats, his Adonaïs, whom he considered to have been murdered by the critics, yet who was now 'made one with Nature', free from 'the contagion of the world's slow stain'. Shelley's thoughts sweep across these pages, the way a crosswind sweeps over a calm sea. Animated with cosmic signs and wonder-animals, from swallows to 'obscene ravens', careering to its un-ending, the book bursts, images pulling at their binding, unstitched with violent imagining.

On the cover of another journal, Shelley had used a knife to carve out the words SINCERITY AND ZEAL. Here, almost every word or line is crossed out in favour of a superior idea or arrangement; and that in turn revised, revision upon revision, cut up and sliced. The poet's thought processes visibly arrange themselves across the pages, even as they fade away and trail into nothingness. He was racing his own fate, this starman. Sometimes the words seem to gather in an unfilterable cluster, heading down a dead end. Momentarily defeated, or perhaps rocked by a wave on Don Juan, Shelley stops, overwhelmed, and starts to doodle a tree or a boat, his quill quivering over the page. With the words 'eternal living stars, which smile on its despair', he pauses to add a stellar mass, an exploding future vision of a black hole.

Now other images hove into view, scratched into the thick Italian paper: a prancing nude cartoon man; a huge phallus. And strangest of all, a spiky demon, a cousin of that Tan-yr-allt imp, morphing out of an inkblot into an angular semi-human figure with talons and a bird's beak, breaking free from a tree or out of a painting by Bosch. Even unto the last, as the waves rose up and he was sucked under, Shelley remained a haunted man.

The breath whose might I have invok'd in song
Descends on me; my spirit's bark is driven,
Far from the shore, far from the trembling throng
Whose sails were never to the tempest given;
The massy earth and spherèd skies are riven!
I am born darkly, fearfully, afar;
Whilst, burning through the inmost veil of Heaven,
The soul of Adonaïs, like a star,
Beacons from the abode where the Eternal are.

ZEROANDEVERYTHINGTOGETHER

I lie in the ward through the night, watching wild people in their beds, caught in their uncurtained dramas. A ninety-two-year-old man cries out for someone to come and relieve the pain in his neuropathic legs, when he has already been told there's nothing they can do for him. His reiterated groans are awful to hear, but you get used to them after a while, you even switch off, like the nurses do as they get on with their jobs, logging our data onto the computer, waking us up at regular intervals to shine torches in our eyes and ask if we know where we are. I don't even know what floor of the building I'm on. Am I on the top or down in the basement? It could be day or night or neither in here. There are no windows to betray the natural world; only the hermetic world of medicine, a sealed culture unto itself, responsible to unspoken codes designed to perpetuate as much as to minister.

When it first happened, time stood still. Or perhaps it went into reverse, rewinding to wipe the precise moment of impact from my memory, even before I'd experienced it. The progress of my life stopped; and restarted. Whatever occurred in between was lost, dissolved by the adrenaline speeding up my heart to hasten me out of the trauma. Entire spaces fell out of my brain. I knew I had a good friend in the same city, but I couldn't remember her name or even her face. My eyes swam. Yet I felt no pain, standing there over the dark pool that my head had left behind in the gutter as a Polish woman, the first on the scene, said, 'Oh no oh no oh no.' I asked urgently for water, and when it came I used it to try and wash the blood out of my favourite stripy shirt.

Later, an investigator would examine the CCTV footage of the accident, a playback in grainy slow-motion. Hurtling through a kind of tunnel created by a pair of tall buildings, I do not notice the low barrier slung across the road. My new brakes are all too efficient. I see the obstruction too late, but not too late to brake. And as I do, my bike is left behind and I carry on, tumbling like an acrobat over the unlit barrier and landing on my head, hitting the tarmac with my brow.

The fact that all this takes place in twilight only makes it more cinematic – and not just in retrospect or in the film I'll never see, but in the unscheduled performance outside which I stand, stranded, pulling myself to my feet and feeling only annoyance that I've missed my train home and worrying about my bike lying abandoned on the kerb. I want to see it all rewound, from another angle, in the way that I witnessed my first accident, standing at the bus stop under the butcher's clock on the way to school one morning, when a boy and his bike collided with a car at the busy junction, his body flying through the air, tumbling over the bonnet with a bounce as he shouted out 'Mummy,' much as deserters called for their mothers when they were shot at dawn.

The darkness of the night seemed to swallow up what followed. I don't remember how I got to the hospital, although I do remember a security guard by my side in the car and wanting to be held, to stop the momentum of the crash still reverberating through my body. I felt utterly abandoned and alone. As I waited to be seen in A&E, the geriatric woman opposite me sat up in bed, having taken off her blouse – it was one of the hottest nights of the year, in a heatwave that had lasted for weeks – exposing her yellowy breasts. She too was calling. 'Why won't someone come and help me?' And I thought of my mother in intensive care in the last week of her life, wired-up in a darkened airless ward, the instruments around her beeping out the fathoms of her descent as she sank into some unknown abyss. A young nurse stitched up my eyebrow, sewing it back together. I wished she wasn't chatting as she remade my face.

Wheeled into the ward for overnight observation, everything is twilit, warm and womb-like. It is comforting to be among strangers. During the night I get up to piss, but almost faint as I do, a result of my low blood pressure, compressed in this bathysphere. The doctor orders that I should be hydrated. A handsome nurse tries to stick a needle in my thin arms. Unable to find a vein, he repeats the procedure again and again, digging around in my skin. I apologise. Then I feel the cold flow of the saline, the sea inside of me.

'There's not much of you, is there?' says Sue the ward nurse, altogether too breezy for 3 a.m.; she talks and laughs out loud with little concession to the hour or her sleeping patients. Lying in pain I can't quite discern, I do not hold it against her. Three hours later the light levels shift, as they do at the end of an overnight flight, and I receive two slices of toast, my reward for having survived.

All the rhythms of this place are removed from the world. It is a surrogate society to which anyone might belong yet where no one really does. It offers a kind of physical transcendence, a halfway house; a state suspended between bleeping machines and human bureaucracy, where bodies are more important than people. As a boy in the nineteen-sixties I feared hospitalisation almost as much as I feared conscription, sucked into soulless organisations bent on my suppression. I remember visiting my father in an old brick hospital on the other side of the city and seeing him lying in the men's ward. The room was gloomy and oppressive. He was suffering from a suspected heart condition; he told us how one of his fellow patients had blister burns the size of tomatoes. The redness and roundness and the ripeness lodged in an interior from which one might never escape; although now I wonder if I ever saw that scene, instead of imagining it from our car parked outside.

In the morning, I'm wheeled out again, not into the light, but into the darkened x-ray room, where I'm spread out on a table, unetherised. There I'm examined for breakages and whatever else might have happened to all that stuff crammed inside my skin. My ribcage is briefly irradiated; I think of it lit up inside me. Then my arm is stretched out on a leaded mat marked out with white lines like a miniature tennis court. I'm being squared up and sectioned. The radiographer delivers the news sympathetically: my left hand, with which I write, is fractured, and will need a cast.

When you are young, you are conscious of your body's perfection; the little god that you are. The older you get, the stranger your body becomes; a covert, skin-covered landscape of yourself, supported by your own scaffolding. I remember something Mark said about how your skeleton falls into place as you get older, getting comfy in its positions; he wriggled his shoulders as he told me this. I fall in with a pattern set by myself and my parents; I might or might not be recognised from my interior as much as anyone might or might not know me from outside. You know your own body as much as you know the wiring in your house. We all carry our selves as if we were burdens rather than miracles.

And what seems so familiar is nothing of the kind. I was once told that my shoulderblades were unnaturally far apart; that one shoulder was inordinately developed compared to the other; that one leg was longer than the other, too, and that my spinal column was bent in a sort of S shape. And complaining of short-ness of breath after a long cold winter, I saw another doctor type on her screen that I had 'air hunger', as though I was greedy to live, taking up other people's oxygen. So I wasn't the same person I thought I'd been, all these years; I was as different as we all feel we are. After all, I've had at least six or seven skeletons in my life, my bones continually replaced. Like you I am a wonder of regeneration. Perhaps I might yet metamorphose and my scapulae grow those wings; or maybe being so strangely formed accounts for why I feel happier in the water – if only I could breathe there, too.

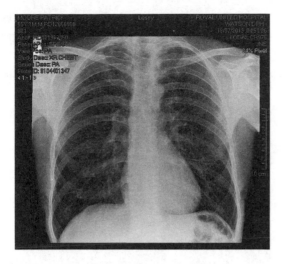

Later, I'll open the envelope I'm given by the radiographer and look at the images illicitly, as if spying on myself. Sliding the disk into my white computer, I see my inner space: the shadowy me, the me-ness of me no one can see, not even me. It might be a photograph of my soul, or a sea creature with a wonky spine, all flickering and glowing, all clouds and bones, skeins and roots, growing into ghostly coral and pearls.

Back home I photograph myself in the mirror, through another lens. My pale eye looks out through the lurid bruises as they ripen, in the same way a piece of fruit ripens and decays. My face changes over the days, sequencing like a time-lapse film as it seeks to repair itself, to return me to me. I look closer at my eyes, as if I'd never seen them before or known what colour they were. I think of my mother's eye, which was not the eye she was born with but a graft, a cornea taken from a young man killed in a motorbike accident in an unsuccessful attempt to treat her glaucoma. It was his eye, stitched into her own, that I looked at when I kissed her goodbye, on another hospital bed, her red hair turned silver, and splayed on her pillow.

I'm left with the altered topography of my body. A new natural history. My wounded eyebrow is restitched with electric-blue sutures, the spreading stain like badly-applied glam make-up, the bruises blushing pink and methylated purple and iodine-yellow like some decadent flower beneath my skin. I yearn with sympathy for my stupidity. And I think, this is my last transformation, the last time I looked beautiful. The first was a long time ago.

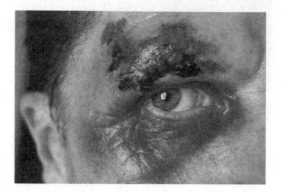

In Edwin Ernest Morgan's glamorous painting from the nineteen-thirties, Beacon Cove is alive with radiant sun-worshippers. They are arrayed along its semi-circular tiers which lead down to the sea like a temple to health. Men in soft white shirts and flannels, their collars upturned to signal their sporty appeal, chat with young women with bobbed hair, wearing tight bathing costumes that show off their lithe bodies and tanned legs. Agatha Christie liked to swim off this Torquay beach as a young girl, although on one occasion she nearly drowned here.

Books are read, children play. Emerging from the half-timbered, green-and-white-painted tearoom, once a lifeboat shed, an aproned waitress with a white headdress, looking a little like a nurse, serves tea to her customers sitting on cane chairs. Others lounge in shaded deckchairs. There's a faint air of the interwar sanatorium; the artist painted his picture from the vantage point of the medicinal baths set on the rocky headland overlooking the cove, an establishment which contained and concentrated the elements of the resort, offering seaweed, sulphur and pine baths, and a Vita Glass Sun Lounge, complete with 'the latest apparatus for administering ultra violet rays'. Like a railway poster or a postcard, continually reprinted to sell the sea, Morgan's painting dates the eternity of a resort; the way it doesn't change and always does. These people are the same people, always, only in different clothes. It doesn't matter that when I visit Beacon Cove in the evening, its only occupants are a group of teenagers doing drugs, looking up at me shiftily from the concrete steps.

We'd set out at four o'clock in the morning, long before the sun rose, the family car fully packed. We were fourteen-night refugees, fleeing the fifty weeks my father worked in a cable factory on reclaimed land by the docks. Halfway there we'd stop in Lyme Regis for tea from a flask and white rolls warm from the bakery, eaten as the car's bonnet creaked with the heat of its exertion, getting its breath back. The sea was beginning to lighten into day. I was still in shorts. My two younger sisters wore matching dresses. When we lined up for photographs, our knees were the most prominent feature: the bones of three thin children, products of a post-war world.

It was only when the earth turned red that we really knew we had left home behind. The car rounded the last hill to reveal the turquoise sea and exotic palms and a place geared not to industrial production, but to the freedom that the water allowed: from time-stamped work, from never-ending housewifery, from the terrors of school. It was all caught up in the sweep of the bay, all there for us: from the posh houses on the far hill, working round the harbour and past fish-and-chip shops and the park to its southern end with its amusement arcades.

Our big old car would climb out of town, past the beaches and up the steep hill, into the gates of the caravan park. It was an alternative, semi-temporary settlement; for all we knew, it only existed for the two weeks we spent here. With its rows of tin caravans moored on the hillside and linked by concrete paths and wash-houses, it hovered between holiday camp and prefab site; another legacy of the war. At the office, we'd collect the key for our caravan and install ourselves in our narrow bunks, above which were little plywood cupboards with sliding doors where we'd store our possessions. We were in a boat moored on the grass.

My mother would prepare an 'emergency' dinner assembled out of tins, a luxury and a necessity of the hour: new potatoes, pale and round, bobbing like definned fish in the pan, and garden peas, olive-green and sweet in soupy liquid, all released by a cow-headed tin-opener brought from home, so hefty it could have opened the caravan itself. Its bovine knob turned corned beef out of a third tin, a grey-pink brick of stringy muscle, fat and corn, ready to be sliced onto our plates. It might have been a meal served in an air-raid shelter.

Our holidays were spent in such intimate spaces; only now do I realise that, born in May, I was probably conceived by the sea.

This was where I grew old and when I stayed the same, since it was only then that we were photographed, caught in the colour of childhood from one summer to the next. We had a dog then, too, or at least my brother did, a wild dark mongrel named Bimbo. He lives only faintly but physically in my memory, as if he were my age. He was run over on holiday, a trauma which I do not remember at all, perhaps because he survived the van that drove over his belly and only died weeks later. He looks up at me from my brother's side. Dogs in photographs are always already dead, lost in time.

Off the top of the wardrobe, in our nineteen-twenties house, I pull a box of transparencies taken by my father on his Instamatic, always housed in its brown canvas case and slung across his shoulder like a military accessory; it was part of his jaunty holiday outfit, along with his blue shorts and his bright red short-sleeved shirt with its print of blue-and-yellow yachts. He liked clothes, as did his tailor father, whose voice Dad echoed when he said, 'Quite the fashion plate,' if I were wearing something new. He detected the pride. But he was a particularly bad photographer.

Every pose of the four of us was erratically framed: my mother in her succession of hairstyles and her own bright holiday clothes, as bright as her red hair, and we three children, my youngest, blonde and freckled sister smiling gappily for the camera through her missing front teeth; my older sister, taller, with her long dark hair, looking more solemnly at the lens; and me, in my grey shorts and long socks and brown crepe-soled sandals and my red-and-white-and-grey-striped shirt with a floppy red collar. I can still feel that shirt now, feel the joy of pulling it over my head. All of these snaps show us with our heads or limbs cropped by Dad's steady finger on the shutter button, clicking through the years: standing on a Dartmoor tor, sitting on the car's bonnet outside our caravan; crouching on a red beach, my skinny shoulders, my shiny green trunks; successively growing apart from the world.

I slot the slides into the clunky viewer. They're Kodachrome windows into my past, the holiday sun still shining through them;

each a cell of slowed-down time, scenes from a documentary no one ever bothered to make. The sliced-up suburban movies synopsise my history in an edited sequence: from my smiley face sitting in a fold-up chair, to my sulky teen scowl. My badly-cut hair grows longer, directed by my hormones; my legs grow longer too, out of shorts and into trousers. My clothes are no longer a different-coloured version of my school uniform. One slide shows me in white shirt and wide white jeans on a campsite outside Paris, the only time we went abroad; a pair of teenage French girls said to me, 'Vous êtes un homme de L'Orange Méchanique?' Holidays take on a new significance; songs on the car radio acquired a new importance as we pulled onto the promenade in the early morning, the seventies light flaring on the scene as a sleek performer with hair so black it looked blue sang about being in with the in crowd. The windblown cordylines and peeling stucco of the English Riviera melded into the Côte d'Azur or Bel Air, and I dreamed of blue pools and white tuxedos and invitations from famous writers in the nineteen-twenties.

At the caravan park's pool – a concrete tank surrounded, not by blondes in long dresses and art-deco palms, nor even filled with performing dolphins and an abandoned orca, but lined with pink and yellow paving slabs, plastic chairs and an ice-cream kiosk; an intimidating place to a boy who could not swim – I'd loiter in the shop, looking along the rows of holiday reading. Beneath the sunned spines of romantic novels were the publications that fed my own dreams: double-page posters which opened up into pin-ups of pop stars in black sequins and slick quiffs, or a god in a powder-blue suit and eye make-up to match, set next to women's magazines with artfully photographed fashion spreads. In one, inspired by that year's film of The Great Gatsby, cloche-hatted models in ivory satin clutching long cigarette-holders were escort-ed by languid young men in white linen suits, their golden hair swept back immaculately. They strolled in an airy limbo, these careless people, to the half-mocking strains of 'What'll I Do?' and 'Ain't We Got Fun?'.

I cannot express what those images stirred in me. Something beyond desire. Something compulsive. Something far removed from my surroundings and yet part of them, too. They still make me ache.

In the summer of 1838 a young woman arrived in Torquay, sent there by her father to alleviate an illness no one could name, but that had rendered her an invalid since she was a teenager. For her this seaside resort was just that: a resort, perhaps her last.

Elizabeth's family, the Barrett-Moultons, owned slave-run plantations in Jamaica. Sugar had paid for the house in which she grew up, an ornate, oriental-styled mansion in Herefordshire resembling Brighton Pavilion, complete with concrete minarets, doors inlaid with mother-of-pearl and a tunnel that led from the house to the gardens and their grottoes. Ominously named Hope End, its fantasy was bought by the toil of stolen people; it might have been built out of sugar. Set against its refined whiteness, Elizabeth believed herself to possess black blood, a result of her ancestors' liaisons.

At fifteen, Elizabeth begun to suffer chronic ill health. She took opium to relieve her symptoms, and would do so for the rest of her life. But her true solace was poetry. She was already writing and publishing her work while Keats, Shelley and Byron were still alive, and she fell under their influence; the onset of her illness coincided with the appearance of Shelley's *Adonaïs*. She was, in her mind, heir to Shelley, fated with his spirit, as if with his watery death his power were to pass on to her. In her fragile state she held fast to his 'perfect exquisition'. It was her article of faith, her destiny. 'I always imagine,' she wrote, 'I was sent on the earth for some purpose.'

And as with the contrast between Shelley's physical body and his ferocious spirit, so it is difficult to reconcile Elizabeth's fierce ambition with the shy face that peers out from her dark curls, their weight making up for the weakness of her bird-like body. No one was ever sure of the nature of her suffering, not even Elizabeth herself. She was subject to that uncertainty. She spent months suspended in a 'spine crib', a hammock hanging four feet off the floor as though she were on a hospital ship. Heated glass cups were used to bring her blood to the surface, leeches applied to suck it out, and, perhaps most extraordinary of all, setons were employed – silk sutures or strips of canvas threaded on needles through pinched folds of her skin, as if she were being sewn up. Cupped, bled, pierced and suspended, her treatments amounted to an exorcism, drawing out bad spirits or ectoplasm. It was as though the century itself, all its appropriations and abuses, was impacting physically on her body, her frail flesh martyred to the Industrial Revolution.

In 1832 a black rebellion in Jamaica and the impending abolition of slavery forced the Barretts to abandon Hope End and its domes. Their fortunes reduced, they moved to Sidmouth on the Devon coast, partly for Elizabeth's health. She was entranced. She declared the sea to be visible poetry, 'the sublimest object in nature', where 'the grandeur is concentrated upon the ocean without deigning to have anything to do with the earth'. She loved the moonlight on the water by night, and her sister Arabella swam in it every day, even in December.

The sea was an opening more effective than any suture. And Elizabeth's happiness was complete when her beloved younger brother Edward came back from the family's estate in Jamaica, his return made more dramatic when he'd 'nearly died a glorious death' on the way home across the Atlantic, supposedly poisoned by '*a dolphin which had hung in the moonshine!*' It was a vodou-like notion, evoking the watery spirits of the Caribbean, where I once saw a boat rowed out with a Haitian mambo priestess sitting in it, her head wound with a white turban, casting offerings of cigarettes and rum to the *lwa*, the spirits of the deep who were thought to drag the unwitting down into their domain. 'The moonshine poisoned the dolphin, and the dolphin poisoned Bro,' Elizabeth reported, unaware that in the Caribbean, the dolphin was also a fish, 'and poor Brozie grew quite black and swollen in the face. Would it not have been a glorious death – to die of a dolphin and moonshine?'

After three years in Sidmouth the Barretts resettled in Marylebone, London, in whose airless streets Elizabeth's health relapsed. Suffering from affected lungs, it was decided that she must return to the sea, and so she was sent to Torquay, where her aunt and uncle lived. Unable to bear the motion of a carriage – one modern diagnosis of her condition suggests that she suffered from spinal tuberculosis, causing extreme sensitivity and twisting to her vertebrae – Elizabeth sailed from London to Plymouth, along with Edward and two of her sisters. She feared the journey more than going to the North Pole, but she was the only woman on board not to suffer seasickness, and boasted of her 'oceanic reputation'. After staying with their relatives, Elizabeth and her siblings moved to a tall townhouse on Beacon Hill, overlooking a signalling point on the harbour at one end of the great curving bay. At the other, in the distance, was Berry Head, in whose deep waters dolphins swam and gannets dived and where, in 1815, sightseers had flocked to catch a glimpse, not of a whale, but of the Emperor Napoleon on HMS *Bellerophon*, on his way to exile on St Helena.

Torquay was newly fashionable; Tennyson, who visited in the same year that Elizabeth arrived, called it 'the loveliest sea village in England'. Backed by the wildness of Dartmoor, set between the Dart and the Teign, its climate was dry, and its hills strewn with white villas and terraces. It was as close as England could get to the Mediterranean. The blue sea clashed with the red cliffs, scorched when prehistoric Devon had lain under equatorial sun. At last, Elizabeth was back by the sea. 'Here, we are immediately upon the lovely bay,' she wrote in a letter, 'a few paces dividing our door from its waves.' Their windows looked onto the water, 'and our ears are as familiar as its rocks are, with the sound'. Like her hero Shelley at the Casa Magni, she was in direct communion with her muse.

'Our house here is in the sea,' she told another friend. 'At least to my imagination it is – which is the same you know as its being so actually.' Whenever the steam packet entered or left the harbour her bed shook; the building at the end of their terrace was actually flooded by the sea, by invitation. The Bath House, established in 1817 during the Regency craze for sea bathing, allowed the tide to flow 'through the wall of the pier into a spacious reservoir', where it was filtered to remove weed and other organisms. In this therapeutic machine, a human aquarium, Elizabeth could take the salt-water cure without having to brave the open sea. Yet she was determined to experience the water, and despite her frailty she often went sailing, accompanied by her maid Crow, perhaps on loan from Nightmare Abbey.

Suspended from the land without medical aids, Elizabeth discovered a new vision of delight. 'My love of water concentrates itself in the boat,' she said. But there were plans to upset her idyll. When her aunt and uncle moved to Merry Oak, on the outskirts of Southampton – half a mile from where I am writing this – it was proposed that Elizabeth should go with them. She was horrified. It may have been a fashionable spa, but Elizabeth had 'taken a great dislike for Southampton … on account of the dampness of the place, occasioned by all the wet mud of the river', even though she had yet to visit the town. Her physician, Dr Barry, supported her prejudice, and expressly advised against the move as potentially injurious to her health. Just as Sylvia Plath was never sent to Provincetown, Elizabeth did not go to Southampton; she was allowed to remain in Torquay – despite the fact that, 'as to its human aspect, it is much more like a hospital than anything else'.

The sea's promise of life was undermined by the presence of death, here at the termination of the land. And although the water was always in view, Elizabeth was confined, like those sea baths; sometimes she would not leave her room for weeks on end. Her days were only relieved by the presence of Bro, who would bring her presents such as a 'very beautiful silver remember medal' of Byron, and who lay talking on her bed for hours. She may have disapproved of him attending parties at which laughing gas was inhaled – worried that, at the age of thirty-one, his talents were being dissipated in a seaside place, where life seemed so much looser – but Bro was more like a lover to her, filling the space a suitor might have occupied.

Despite Bro's attentions, Elizabeth became bored and depressed. The household moved to 1 Beacon Terrace, a cheaper lease. She had lost faith in the sea, relying on a different drug. 'Opium – opium – night after night!' she wrote, 'and some nights even opium won't do.' Her lungs began to haemorrhage; her skin was constantly blistered by her doctor. After two years in Torquay her health did not seem to be improving at all. Rather, the reverse: she was convinced that death was near.

As it was: but not hers. In February 1840, Elizabeth received news that her younger brother Sam had died of a fever in Jamaica. She was still recovering from the shock when five months later, on 11 July, Bro, along with his twenty-one-year-old friend the Honourable Charles Vanneck, went sailing on *La Belle Sauvage*. Also on board was the thirty-five-year-old Captain Carlyle Clarke of the Indian Army, and an experienced young sailor named White. None of them returned alive.

The yacht was last seen that Saturday afternoon, two or three miles off Teignmouth. It was a summer sea with only a brief squall. The witness, the aptly named Richard Wake of Heavitree, observed that the boat was 'sailing under a heavy press of canvas at the time she went down'. She seemed to sink in seconds, as if a hand had reached up from below. One of Wake's sailors saw the mast sticking out of the water – all but with a cormorant perched on its top – 'and that disappeared'. When Wake arrived on the scene there was no trace of the boat or its crew; 'not a hat, oar, or any other vestige remained'. Boats sent to drag the bottom with grapnels found nothing either.

As Mary Shelley had awaited news of her husband, so this lack of evidence created a terrible absence. It was as if Bro had been taken by the spirits of the sea. The cruelness lay in the water which Elizabeth could see from her window: 'he had left me! gone! For three days we waited – and I hoped while I could – oh – that awful agony of three days! And the sun shone as it shines to-day, and there was no more wind than now; and the sea under the windows was like this paper for smoothness – and my sisters drew the curtains back that I might see for myself how smooth the sea was, and how it could hurt nobody – and other boats came back one by one.'

All in the party could swim, save the sailor. Perhaps Bro had made it ashore? Their father, arriving to deal with the disaster, indulged in the hope – 'billow upon billow pass over me' – that his son might be found alive. But the local paper was hardly encouraging – 'Fatal Catastrophe off our Coast! Four Lives Lost!' – quoting, with further melodrama, from William Falconer's 'The Shipwreck', 'The hostile waters closed around their head | They sank, for ever numbered with the dead', and publicising rewards for anyone who found linen cast ashore 'marked with the initials of the beloved dead'.

Six days later, Captain Clarke's corpse was pulled from the water by a Brixham trawler crewed by the two sons of the fishing port's vicar, Francis Lyte, the composer of the hymn 'Abide With Me'. Clarke's body was unblemished, his buttonhole still in his lapel. It took three more weeks before Bro's body was seen, floating a mile and a half out in the bay. 'On being examined it was found a little mutilated in the face and hands.' His gold watch, a purse containing sixteen shillings in silver, a pencil case, cigar box, gold ring and pocket handkerchief marked E.M.B. were still about his person: all the accoutrements of a carefree young man about town.

To lose a brother and a son dismantles a family, throws the future into doubt. At the age of twenty-three, my brother crashed his car and lay comatose in hospital for a week. We too experienced days when he was neither with us nor we without him. I watched through the banisters as my father wept, an awful sight; and at the age of eleven, I refused to accept that Andrew was never coming back, partly because I still saw his fuzzy shape outlined in my bedroom doorway, like an after-image from the sun.

Elizabeth was racked with guilt, because her father had disapproved of Edward going to Torquay. She was being punished. The sea which had summoned her in hope had taken her hope, the one person for whom her life was worth living. Now she could not wait to get away from 'this dreadful dreadful place … These walls – & the sound of what is very fearful a few yards from them – that perpetual dashing sound, have preyed on me. I have been crushed, trodden down. God's will is terrible!'

Later, after she had become the most famous female poet of her age, legend would create an impossible scene of Elizabeth on a balcony, looking out to the bay as Bro perished in the waves, as if she had watched him being dragged down below. 'The associations of this place, lie upon me, struggle as I may, like the oppression of a perpetual night-mare. It is an instinct of self-preservation which impels me to escape – or to try to escape.' The sea evoked weird, woozy images that seemed to be products of her addiction. In an essay on the Greek poets published soon after Bro had drowned she discussed a dead language, 'pang by pang, each with a dolphin colour – yielding reluctantly to that doom of death and silence which must come at last to the speaker and the speech. Wonderful it is to look back fathoms down the great past, thousands of years away.' Transported in time and space as Virginia Woolf would be, she too associated her loved one with a cetacean: 'Faint and dim | His spirits seemed to sink in him – | Then, like a dolphin, change and swim.'

As soon as she could, Elizabeth left the sea behind. On 1 September 1841 she was taken from Torquay in a specially sprung carriage with a bed to allow her to remain supine; it took ten days to reach London as they had to stop frequently to allow her to rest. The sound of the sea was now hateful to her; she retreated to her dark bedroom deep in the city to drown out the noise. 'I cannot look back to any month or week of that year without horror, & a feeling of the wandering of the senses,' she would recall. 'Places are ideas, and ideas can madden or kill.' She told John Ruskin, 'I belong to a family of West Indian slaveholders, and if I believed in curses, I should be afraid.' She would give anything 'to own some purer lineage than that of the blood of the slave! Cursed are we from generation to generation!' It was as if she were labouring under a vodou hex; as though Bro, heir to those estates, had died under a malediction.

Elizabeth's own life had ended; her illness would never be dismissed. All she faced was a reduced existence. Dressed in black silk, her tiny pale-dark face and huge eyes and strangely wide mouth framed by her thick brown curls – 'Of delicate features, – paler, near as grave' – she seemed unnaturally preserved by grief and seclusion, all but tattered and frayed. Immured in her mourning room, she lived on 'obstinacy and dry toast'. She was thirty-three but seemed more like a teenager; a modern doctor might diagnose an eating disorder, another kind of dysfunctional consumption, an absolute sensitivity to everything. 'The truth is I am made of paper and it tears me,' she said. (A century later, Woolf, who became fascinated by Elizabeth, would claim, 'Cut me anywhere, and I bled too profusely.') Her only companion now was Flush, a red cocker spaniel given to her in consolation for Bro's loss. He was as loyal as Byron's Boatswain and constantly at her side – save when he was three times snatched in the street by dognappers, and Elizabeth had to brave the terrors of Shoreditch to retrieve him. Perhaps he was Bro reincarnate.

They looked extraordinarily alike, these two, poet and dog, with their silky ringlets and expressive eyes. She imagined Flush 'as hairy as Faunus', and thanked 'the true PAN, | Who, by low creatures, leads to heights of love'. A century later, in her biography of Flush, Woolf imagined the poet in her bedroom, the dog's face and bright eyes close to Elizabeth's: 'Was she no longer an invalid in Wimpole Street, but a Greek nymph in some dim grove in Arcady?' But Flush, who had once hunted rabbits in the countryside, was reduced to lying in a patch of light that moved across the carpet as the sun swung over the city, day after day.

Woolf would claim that years of reclusion had done irreparable harm to Elizabeth as an artist; that she had been somehow maimed. As an invalid she was rendered invalid; like other Victorian women, she responded to constraints with withdrawal. Yet in her immobility she lay open to the world. As Virginia sent her Imagination over the ocean, and as Stephen travelled in his reveries, so Elizabeth reached out of her window like Odin's ravens to bring back news of the world beyond. 'Religious hermits, when they care to see visions, do it better they all say, through fasting and flagellation and seclusion in dark places,' she noted.

With her morbid intensity – 'I am Cassandra, you know, and smell the slaughter in the bathroom' – influenced by her fellow addict, Coleridge, and influencing in turn Edgar Allan Poe (who dedicated 'The Raven' to her), Elizabeth became a mythic figure. Confined, she rejected convention; her letters and poems burst with passion and outrage, conjuring up images and visions and self-drama. Restricted by reality, she looked to the other: she became fascinated by the occult, by spiritualism and animal magnetism, while looking like a black-clad Miss Havisham or Kate Bush in a crinoline. In another age she might have stood accused of being a witch, with Flush as her familiar.

In her bitter poem of 1848, 'The Runaway Slave at Pilgrim's Point', a strange meeting of New England and the Caribbean, Elizabeth imagined a young black woman standing on the shore of Cape Cod Bay, listening to 'the ocean's roar' and cursing the white land. Raped by a gang of slave-owners, she suffocates her bastard child for its whiteness and evokes vodou lwa: 'Your fine white angels ... sucked the soul of that child of mine.' There was, as Elizabeth admitted, a 'deathly odour' to her work. Its power lay in her protest at modern evils: slavery, child labour, and the enshacklement of her sex. The future seemed to offer only bitterness.

But on Saturday, 12 September 1846, all her expectations were overturned. Aged forty, she defied her father and secretly married Robert Browning, scion of another slave-owning family, himself part Creole. The following Saturday, Elizabeth snuck out of the house like a teenager with Flush in her arms and her new maid Wilson at her side. She met Browning at a nearby bookshop, then took the five o'clock train from Vauxhall to Southampton. They sailed at nine o'clock that night from the Royal Pier on the South-Western Steam Packet Company's 'Splendid and Powerful STEAM SHIP' to Havre-de-Grâce, 'main cabin, twenty-one shillings, dogs, five shillings'.

Robert worried that the evening voyage would present new dangers. On board, Elizabeth and Wilson loosened their stays, enduring every pitch and roll. Flush, too, suffered. Despite his five-shilling ticket he was turned out and treated, well, like a dog, 'inasmuch as people have a barbarous mania for chaining dogs upon deck all night'. Defiantly, Elizabeth brought him into the cabin, but a woman with six screaming children object-ed, 'and delivered him over to the tormentors though he had escaped from them to me for the third time'. Weathering the storm, the exhausted refugees – two newlyweds, one maid and one dog – made it across the Channel. No one came in pursuit. Elizabeth's father never forgave her; he returned all her letters unopened, even those bordered in black. She was already dead to him.

As Mrs Browning – an oddly ordinary title, given its extraordinary achievement – this black-clad butterfly would open up in Italy, in the way her heroes had done. It was as if she had been born for this. She basked in the sensual heat, wearing the loosest of gowns. In 1849, aged forty-three, Elizabeth gave birth to a son, whom she called Pen, a neutral name. She would keep his long hair in ringlets and dress him in romantic clothes, black velvet tunics and breeches, declining to define his gender – 'If you put him into a coat and waistcoat forthwith he only would look like a small angel travestied' – and told her brothers not to instruct him in manly ways. As he grew up, Pen came to resemble Bro, too; perhaps that was why Elizabeth disliked the idea of him learning to swim.

This trio had a faerie, androgynous air about them; even Robert 'resembled a girl drest in boy's clothes', according to Elizabeth's friend Miss Mitford. 'He had long ringlets & no neckcloth ... Femmelette – is a word made for him.' Although Elizabeth often appeared pale, both she and Robert were described as having dark complexions: Elizabeth herself said 'I am small and black.' William Michael Rossetti said they took up almost no room in a railway carriage, and barely needed a double bed at an inn. All three seemed out of time and space, sex and race. 'When I look in the glass,' Elizabeth told a friend, 'I see nothing but a perfectly white & black face, the eyes being obliterated by large blots of blackness.' To Nathaniel Hawthorne she was 'that pale, small person, scarcely embodied at all ... her black ringlets cluster down into her neck, and make her face look white by their sable profusion'. Hawthorne, whose interest in the supernatural reached back to the Salem witches whom his own ancestor had tried, regarded her as one might examine an animal or a myth: 'I could not form any judgement about her age, it may range anywhere within the limits of human or elfin life.' Meanwhile, the thrice-ransomed Flush ran wild and learned to speak Italian, according to his mistress, and took to the water with ease, having been baptised in the river 'in Petrarch's name'.

Out of Europe's revolutions came *Aurora Leigh*, a sprawling poem with its heroine based on Elizabeth, and its hero, Romney Leigh, on her brother. Intended to be 'intensely modern', as she believed herself to be, she called it an 'art-novel', its power drawing on the same stormy mid-century disruptions that would influence Emily Brontë and Herman Melville. The poem explodes with Promethean myth and Shelleyan imagery. It cites *The Tempest*, New England utopias and the Irish Famine, and invokes Tahitian queens and Haitian presidents, set next to science-fiction scenes of stars as 'overburned' suns, 'swallowing up | Like wax the azure spaces'. Underlayered by 'marine sub-transatlantic railroads' and fossil mastodons, this is a fearsome vision of a brave new world: Elizabeth had heard accounts of Brook Farm from Hawthorne and her American friend Margaret Fuller, although she disapproved of its socialism, saying she'd rather live under Tsarist Russia 'than in a Fourier-machine, with my individuality sucked out of me by a social air-pump'.

Hyped-up and heretical, *Aurora Leigh* fed on the Hungry Forties, as starvation, revolution and disease swept across Europe, defying the industrial, capitalist age. Aurora is accused of writing 'of factories and of slaves, as if | Your father were a negro, and your son | A spinner in the mills. All's yours and you, – | All, coloured with your blood, or otherwise | Just nothing to you ...' All the while, the sea flows through the blank verse – 'the bitter sea | Inexorably pushed between us both' – with heretical images of shipwreck: '... some hard swimming through | The deeps – I lost breath in my soul sometimes | And cried "God save me if there's any God."' In fact, the manuscript of *Aurora Leigh* was itself nearly lost at sea, going missing as the Brownings sailed to Marseilles, although Elizabeth claimed, flippantly, to be more concerned at the loss of her son's clothes in the same trunk, 'all my Penini's pretty dresses, embroidered trousers, collars, everything I have been collecting to make him look nice in ...'

Her poem may be almost undecipherable now without resort to historical notes, but it had an electrifying effect on her peers. The Pre-Raphaelites worshipped it as an alternative Bible, and saw its author as a kind of patron saint; Ruskin believed it as good as Shakespeare's sonnets; and his pupil Oscar Wilde declared it 'much the greatest work in our literature', 'simply "*intense*" in every way'.

That intensity came out of all those years in her upper room, 'a sort of cage-bird life, born in a cage' – a telling image, given that the presiding spirit of *Aurora Leigh* is an eldritch bird, its spread wings and open beak like Poe's raven: 'We are sepulchred alive in this close world,' Elizabeth wrote. The opium to which she was addicted could induce synaesthesia, bending the senses, hearing colour, seeing music; it once sent me floating down the streets of south-west London by the river, visions flaring out of the brown suburbia around me. But in all her phantasmagoria, Elizabeth remained haunted by the water which had claimed her brother. And in 1850 that memory was stirred up by a new loss and another shipwreck.

She had met Margaret Fuller in Florence, and the two women became close; Fuller, a Transcendentalist and feminist, inspired the utopian passages in *Aurora Leigh*. About to sail back to New York with her husband, the Marchese Ossoli, and their young child, Margaret spent her last night with the Brownings. 'She said with her peculiar smile that "the ship was called the Elizabeth, & she accepted it as a good omen – though a prediction had been made to her husband that the sea wd be fatal to him".'

Elizabeth was nearing New York when, weighed down by a cargo of Carrara marble, she foundered in a storm, only a hundred yards off Fire Island. In the early hours of the morning her freight smashed through her sides. As daylight came, the waves ran too high to launch a rescue – although the land sharks had already arrived with carts to salvage the luxuries being washed ashore. 'At flood-tide, about half past three o'clock, when the ship broke up entirely, they came out of the forecastle, and Margaret sat with her back to the foremast, with her hands on her knees,' Thoreau reported. 'A great wave came and washed her aft. The steward had just before taken her child and started for shore. Both were drowned.' The philosophical receiver of wrecks had been sent by Fuller's friend, Emerson, to search the shore, but he found neither her body nor the revolutionary manuscripts she'd been carrying; no sodden legacy lying, like Shelley's notebook or Prospero's books among the splintered timber and white marble.

Elizabeth received this news having suffered the last and most nearly fatal of four miscarriages. Her reaction was typically intense: 'In sight of shore, of the home, American shore! Oh Great God, how terrible are Thy judgements! The whole associations have been more poignant to me that by a like tragedy I lost once the happiness of my life . . . the life of my life . . . the colour & fragrance of my soul.' The sea had now taken three people from her – Shelley, Bro and Fuller – 'the sea, that blue end of the world, | That fair scroll finis of a wicked book'.

Yet at the last, this fragile, resilient woman may have been reconciled to the waves. In 1858, three years before she died, Elizabeth went back to Le Havre. 'We have come here to dip *me* in warm sea-water,' she wrote, 'for I have been very weak and unwell of late.' While her husband and son swam every day, she was plunged for five minutes at a time into a hip-bath filled with salt water. She could barely get to the sea itself, let alone look at it.

But then Robert found 'a hole I can creep through to the very shore, without walking many yards . . . and the sea is open and satisfactory'. It might have washed away her grief. 'We bathe & get strength, & sit close to the sea watching the animated ships & the swimming men & women.' Even her son had become a water baby. 'Peni bathes everyday & has learned to swim & swims, & looks like a merman, with his locks floating . . .' In the warmth of the sun, looking back over the Channel, there was a sense of healing and a life yet to come.

Three years later, floating on her own opium cloud, Elizabeth opened her eyes in the hour before dawn and told Robert, 'You did right not to wait – what a fine steamer – how comfortable,' as if she were back in Torquay, before Bro left her; as though, like Thoreau, she were sailing away.

Moments later, she breathed her last word, 'Beautiful.'

Early in the morning, before anyone is up and about, I leave my Torquay guest house for the harbour. I ride along the promenade, past the bandstand where I ate fish and chips last night as couples in deckchairs listened to a brass band while boys skateboarded round the park. I cycle up the hill, past the bird compound, its inmates held under a vast net suspended high over the headland, just as Melville described the rigging of a slaveship as 'hung overhead like three ruinous aviaries'. Facing the black canopy, where the birds circle as if caught underwater, is the house where Elizabeth waited for her brother in her own captivity.

Unsettled by a glimpsed scene I would decline to pay for, I leave the avian prison and carry on, past Beacon Cove and the hotel where Stephen Tennant once stayed, half-believing that he was back in his beloved south of France. The sea is deep and turquoise. In silent black and white, a thirteen-year-old girl walks out of Morgan's painting, climbs the rock and leaps into the void, her body curving as if, for a moment, she might fly upward to the sun rather than plunging down into the waves below. The inter-titles record her feat.

She springs forward, like a bird taking wing, and seems to hover.
Then, she dives amidst a splashing of luminous drops of water.

I ride on, ignoring the girl in the water. The road rises to
another headland, where a house appears to be built into the cliff;
its wooden balcony looks out to sea like a crow's nest. It was here,
in a room named Wonderland, that Wilde composed the letters
which would send him to a prison cell.

In 1878 Babbacombe Cliff was rebuilt for Georgiana Mount-
Temple, aristocratic patron of the Pre-Raphaelites and confidante

of Ruskin, whom she introduced to spiritualism; she employed her own mediums to communicate with the dead and was caricatured in Woolf's *Freshwater* as Lady Raven Mount-Temple. Her statue still stands nearby; fresh flowers regularly appear in its bronze hands, placed there by admirers of her championship of animal rights. In 1892 Georgiana lent the house to her cousin, Constance Wilde, so that she and Oscar could escape London for the winter. As dramatic as Babbacombe Cliff was from the outside, its interior was even more remarkable. It was hung with paintings by Rossetti, tiled by Morris and lit by stained-glass windows by Burne-Jones. This secular chapel was a fitting place from which Wilde would order his veiled *Salome* to be bound in Tyrian purple and lettered in tired silver. And although Oscar complained to his friend Robbie Ross, 'Are there beautiful people in London? Here there are none; everyone is so unfinished,' the sea offered him a sense of escape.

It always had. Wilde was an avid swimmer, and had been since his youth. As a muscular student he'd spent his summers swimming in Dublin Bay (in between reading *Aurora Leigh*), where the rocks of the Forty Foot were famous as a naked bathing place for men and boys – both Joyce and Beckett would swim here too. Oscar felt immortal in the sea, and 'sometimes heretical when good Roman Catholic boys enter the water with little amulets and

crosses round their necks and arms that the good S. Christopher
may hold them up'.

Six years later in 1882, on a tour of America which would make
him an international star, an Apostle of his own Newness – not
least because he arrived in Manhattan in an oversized green over-
coat lined with otter fur and collared in seal, almost as if he were
a marine mammal himself – Oscar had spent days on Fire Island,
newly fashionable as a resort for New Yorkers, where the press
portrayed him in a daring costume, complete with sand shoes.
He was a modern man, looking out over the ocean.

Now, after ten years of fame, he was taking refuge from his
own celebrity in Torquay. With Constance called away, leaving him
to look after their two young sons, he invited Lord Alfred Douglas
to stay. Oscar played at being a decadent headmaster, ordering
champagne and sherry and biscuits for himself and Bosie for
morning break, with compulsory reading in bed for his boys at
night. Tradition has it that he swam here too, in the same cove
from which Elizabeth's brother had set sail.

Despite his happiness at Babbacombe Cliff, Wilde detected a
disturbance in the atmosphere. 'But today the sea is rough, and
there are no dryads in the glen, and the wind cries like a thing

whose heart is broken.' His lover's true nature was emerging, tempestuous and petulant. There were angry shouts, and Douglas packed his bags, pursued to London by Oscar's letter: 'Bosie – you must not make scenes with me – they kill me – they wreck the loveliness of life.' Wilde's life would be wrecked by Douglas, as surely as if he'd been cast up on those rocks below Wonderland.

All of this happened with a speed that belied his own dramas. Two years later, in the summer of 1894, Oscar returned to the coast at Worthing. He was working on *The Importance of Being Earnest*, and spent weeks 'doing nothing but bathing and playwriting'. He was a powerful swimmer, as his son Vyvyan recalled, ploughing through high seas like a shark. Evoking Byron's feats, he swam from a yacht as he sailed from Worthing to Littlehampton, only for a sudden storm to stir 'a fearful sea' out of the pitch dark on the return trip; Oscar boasted to Bosie of being 'Viking-like and daring'. The end of the land allowed the unallowable. In Worthing, Wilde wooed Alphonse Conway, a sixteen-year-old boy who wanted to become a sailor. He asked Oscar to take him to Portsmouth. Instead he was kitted out in a blue serge suit and taken to Brighton, a straw hat on his head. A year later, Alphonse's name would be used against Wilde at the Old Bailey.

In London, facing the inevitable, Oscar was encouraged by his friends to flee before it was too late. Frank Harris had a boat waiting on the river, crewed and ready with a head of steam. Harris said extravagantly, 'In one hour she would be free of the Thames and on the high seas, delightful phrase, eh? – high seas indeed where there is freedom uncontrolled.'

He tried to entice Wilde with the vision. 'You've never seen the mouth of the Thames at night, have you? It's a scene from wonderland; houses like blobs of indigo fencing you in; ships drifting past like black ghosts in the misty air, and the purple sky above never so dark as the river, the river like an oily, opaque serpent gliding with a weird life of its own.'

But the water would not save him. In court, the words which he had written from the privacy of Babbacombe Cliff extolling

Bosie's rose-leaf lips were read aloud to the audience in the public gallery. Sentenced to two years' hard labour, Wilde was taken from Pentonville to Clapham Junction where he was spat at as he stood on the platform, his station of the cross. From there he was conducted to Reading Gaol. As prisoner C.3.3 he was subjected to a new transformation. He sobbed as they cut his hair. He was kept in his cell for twenty-three hours a day, and when he did leave it he was forced, like other prisoners, to wear a cap with a thick veil so that no one could recognise each other. Time had stopped for a man who had defined a new age. Like Woolf, he conducted a transaction with his time, only to be punished for his facility. In his pitiful letter, De Profundis, he told Bosie, 'With us, time does not progress. It revolves.' He had not realised that there was so much suffering in the world, or that he would experience it: 'What lies before me is my past.' In his cell, alone and kept in silence, he had 'a strange longing . . . for the sea, to me no less of a mother than the Earth. It seems to me that we all look at Nature too much, and live with her too little.' While the world thought his nature unnatural, he thought of the Greeks: 'they saw that the sea was for the swimmer'.

When he was released after two years' imprisonment, Wilde left immediately for France, as if he could not bear to spend a single night of freedom in the country which had disowned him. At Berneval, near Dieppe, he shook off his imprisonment by swimming every day – 'breasting the waves, a strong and skilful swimmer', according to his friend Robert Sherard, who wished he could take a photograph of the aesthete, 'to show people in England that there's a man in him'.

Oscar had a beach hut set up in which to undress, and told Robbie Ross that he wanted to build 'a little chalet of plaster and wood walls', in which he might live out his life by the sea. He soon found another chalet, with a wooden balcony overlooking the water. Far from Wonderland, it was as bare as Thoreau's hut, its only decoration a carved Madonna salvaged from a fishing boat. He was happier than he had ever been, writing letters precisely timed and dated – 'Thursday 3 June, 2.45 pm, AD 1897' – but addressed from a 'Latitude and Longitude not marked on the sea', a nowhereness which would have appealed to Melville. Oscar had written in *The Soul of Man Under Socialism*, 'A map of the world that does not include Utopia is not even worth glancing at.' He looked out from that atlas and saw his future in the sea.

With his now substantial bulk, far from the immortal student who'd swum in Dublin Bay with Catholic boys and their amulets and crosses, Oscar looked like an elegant elephant seal, his head held above the waves. He adopted a new identity, as Sebastian Melmoth. Surely he had half his life left to live? He was just forty-three. The opal sea washed his sins away while the gulls were blown about like white flowers. He had found his own utopia. Nearby, he'd discovered a chapel, of Our Lady of Joy – 'It has probably been waiting for me all these purple years of pleasure' – and he thought of the Virgin Mary as the Star of the Sea as he lay in the sea-grass outside, recalling the 'strange beauty' of William Michael Rossetti's poem.

> The sea is in its listless chime,
> Like Time's lapse rendered audible;
> The murmur of the earth's large shell.
> In a sad blueness beyond rhyme
> It ends.

'Yesterday I attended Mass at 10 o'clock and afterwards bathed,' he told Robbie. 'So I went into water without being a pagan. The consequence was that I was not tempted by either sirens or mermaidens, or any of the green-haired following of Glaucus. I really think that this is a remarkable thing. In my Pagan days the sea was always full of Tritons blowing conchs, and other unpleasant things. Now it is quite different.'

But a month later he invited Bosie back, telling him that he had a bathing suit ready for him.

UNDERAGREENSEA

In July 1910, seven decades after Elizabeth Barrett Browning had arrived there and eighteen years after Oscar Wilde had left, a teenaged boy from Shrewsbury spent a summer holiday with his uncle and aunt in Torquay. He knew this place and loved it, particularly for the bathing. That year, however, there was a new excitement to his stay. Two hundred vessels of the Royal Navy's Home Fleet had gathered in Torbay to mark the accession of the new king, George V. Yet the spectacle that day was not confined to the sea. For the first time, boy and king would see an aeroplane soaring into the skies.

It seemed like a good omen for a new reign; an intimation of glamour and technology. But it also promised another kind of future, one to which the young man – Wilfred Owen – would bear witness.

Like Elizabeth, Wilfred had come to join his relatives in the town. But unlike her uncle and aunt, living in a grand house on the hill, John and Ann Taylor lived at 264 Union Street in the commercial centre, between a bank and a pub. The wealthy winter visitors to Torquay were now giving way to more ordinary holidaymakers, and the ground floor of the Taylors' house was a shop where they sold books, magazines and stationery. The building is still there: its wide, double-fronted bay windows, which once showed its wares, now display bridal gowns. Wilfred loved the shop; it felt special, the way things do on holiday. Passing the local newspaper office, he imagined a career as a journalist. He thought about the life of a poet, too, although that seemed even more fantastical. He had discovered that Christabel Coleridge, granddaughter of Samuel Taylor Coleridge, lived close by, and one morning he arrived unannounced on her doorstep. Miss Coleridge and her brother Ernest obliged by signing his edition of their ancestor's poems. Other poets he revered had visited the area too. In 'the azure time of June' of 1816, Percy and Mary had spent a second, secret honeymoon in Torquay in 'one soul of interwoven flame'. And two years later, a year after his visit to Southampton and the Isle of Wight, John Keats had stayed for two rather rainier months at neighbouring Teignmouth.

Wilfred felt connected to these poets; they spoke to him over time. Later, he would take the train to Teignmouth, where high tides and winter storms often threatened to cut off Brunel's coastal railway; enveloped in cloud, Keats said Devon was so wet that it was amphibious. He had come here to look after his brother Tom, who was mortally sick with consumption, the same disease that would kill himself. But it was also here, by the sea, that he completed *Endymion*, just as he had begun it by the sea, on the Isle of Wight (and described the whole process as leaping into the ocean to become more aware of his surroundings). Staring into the windows of the house where his hero had stayed, Owen alarmed the inhabitants, so he walked down to the shore, where Keats had seen 'the wide sea did weave | An untumultuous fringe of silver foam | Along the flat, brown sand | I was at home, | And should have been most happy – but I saw | Too far into the sea . . .'

Back in Torquay Wilfred wrote a sonnet to his patron saint: 'Eternally may sad waves wail his death.' He told his mother, Susan, that he was 'in love with a youth and a dead 'un'; and when he read William Michael Rossetti's biography of the poet, he felt that his hand had been guided 'right into the wound . . . I touched, for one moment the incandescent Heart of Keats.' (In a world in which words could pass on as relics, Rossetti himself had befriended the elderly Trelawny, who gave him a blackened fragment of Shelley's skull taken from the funeral pyre.) But for Wilfred, the worship of dead poets was caught up with the sea and the life it offered; the water was the overwhelming reason to love Torquay on this, his third visit there, now with his brother Harold, four years his junior. 'The whole day. . . centres around the bathing, the most enjoyable we have ever had, I think,' Wilfred told his mother, adding, in his sweetly self-important way, 'It is one of those rare cases where the actuality exceeds, does not fall short of, the expectation.'

He may have been born in landlocked Shropshire, but Wilfred had a strange connection to the sea. His father, Tom, would pretend to be a sailor – although his maritime experience amounted to little more than having once sailed to Bombay as a young man. He now worked as a railway clerk, but in his spare time Tom Owen assumed the stance of a captain, dressing up in a nautical, Gilbert-and-Sullivan manner on his visits to Liverpool docks, where he acted as a volunteer for the Missionary Society. One day he invited four Lascars home to tea; Harold remembered eight bare Indian feet appearing under the family table.

Mourning his imaginary career, Tom Owen invested his hopes in his eldest son; he planned a life at sea for Wilfred from the first.

An early family photograph depicts Tom in a sort of seaman's outfit, with a straw hat and wide white trousers. Balancing the infant Wilfred on his shoulder like a kitbag, he poses like an Edwardian *idea* of a sailor. Harold – who really would go to sea – thought their father looked like Robert Louis Stevenson, with his long, handsome moustache. In another photograph, young Wilfred, dressed in his own white sailor's suit, holds a toy yacht made for him by his father. And later, a little older, posed on a swing in navy blue, he seems already set for the sea, looking off to his own horizon.

At the age of six, Wilfred was taught to swim by his father on evening visits to the local public baths, where he displayed 'a lithe aptitude for the water'. On holiday, Tom Owen insisted that his children should swim in the sea every day, whether the sun shone or not. It was a ritual for him, and became so for his eldest son.

At Tramore, near Waterford in Ireland, where many ships had been wrecked, Tom took the boys into a rough sea. Told by local fishermen to return to the shore, he defied them, recklessly diving into the waves and spouting through his moustache looking like a walrus. Later, the Owens caught a large dogfish which they stored overnight in a shed, only for Wilfred and Harold to find it rearing up in a corner on its tail, its mouth gasping and white; they thought it was walking like a man. The resurrected fish was duly released back to the sea, where it continued to be seen, swimming in the shallows. In another eerie incident, the family went for a walk down a dark, tree-shaded lane. A large animal seemed to be moving in the branches above them. At the end of the path a lake appeared like a mirage, and out of the darkness stepped a threatening figure that confronted their father while the children looked on, shaking with fear. None of this was explained, although the fisherman and his wife in whose cottage they were staying looked at each other strangely and asked the Owens not to tell anyone about what had happened.

In Harold's remembering it seemed that the family were collectively haunted, as families can be. Wilfred embodied that mystery; they all felt he was in some way different, with his solemn

gravity but sometimes wildly high spirits. Dark-haired and dark-eyed, he was proud of his Celtic blood, and had an animal love of loneliness. They called him the old wolf, although he might as easily have been a selkie.

Everywhere he could, he swam in the sea. 'We bathe here every day,' Wilfred had written from Cornwall in the summer of 1906. Now a teenager, the water released a new energy in him. Staying in Bournemouth, he told his mother, 'I have been so often at the Sea Side in Day Dreams of late.' Those dreams took shape in a long and enthusiastic poem based on Hans Christian Andersen's 'The Little Mermaid', in which the heroine gives up the sea for the land so that she can meet a young prince; Andersen's story was a cipher, as Elspeth Probyn would write, 'for an impossible love'. Wilfred was influenced by Coleridge's *Rime of the Ancient Mariner* and by Turner, whose paintings he had recently discovered. In lines neatly written out in his exercise book in his open, fair hand, he imagines the 'exceeding deep' beyond the 'sea-bird's view', and 'the tireless glee | Of waggish dolphins turning somersault | And whales a-snorting fountains angrily'. Such scenes make me wonder if Wilfred had seen cetaceans as a child, but his animals seem to have surfaced from *Endymion*, whose 'gulphing' whales and 'Ionian shoals of dolphins' would be familiar to Orlando and Oscar Wilde, too.

That summer in Torquay, Wilfred and Harold found a secluded cove, away from the other beaches with their deckchairs and families staking claim on their bit of England in the sand. To get there the boys walked up the hill, past the house from which Elizabeth had looked out to sea, and over the headland to Meadfoot. Here Charles Darwin had convalesced; his neighbour was Angela Burdett-Coutts, the richest woman in England; another villa was owned by the Romanoffs. The area's well-to-do air appealed to the snob in Wilfred. But beneath the cliffs was the winding darkness of Kents Cavern, where three species of humans had lived; its prehistory appealed to the archaeologist in him. The beach was clean and wide, with flat pebbles veined with quartz; it looked out to a jagged shard named Shag Rock after its sentinel birds and driven at an angle into the sea, like something that had fallen from the sky.

One morning after swimming, Wilfred proposed an impromptu investigation of the geology of the cliff. As the brothers busied themselves with their excavations, as if looking for a psammead, they noticed two boys and a girl also digging about the rocks. Wilfred found the newcomers interesting – especially one of the boys, Russell Story Tarr. His father was Ralph Stockman Tarr, from Gloucester, Massachusetts; his ancestors included mariners and fish-oil merchants and women accused of witchcraft. A renowned geologist, student of marine zoology, collector of meteorites and Arctic explorer, Ralph Tarr was about to leave for Spitzbergen to investigate the physical properties of ice. The Tarrs were rich, evidently: Russell and his sister Catherine were staying with their parents at the grand Osborne Hotel, overlooking the beach.

Russell, bespectacled and about to go up to Harvard, was the personification of a new world. He and Owen were the same age, seventeen, and shared a love of geology; more importantly, Wilfred admired Russell's prowess in the water. 'This American boy is a splendid swimmer,' he told his father. Russell would dive from a moored raft in the bay, disappearing for a dangerously long time, to emerge exhausted but clutching two handfuls of pebbles to prove how deep he had been. Wilfred and Harold tried to emulate him, but came nowhere near to doing so. Besotted by their new friend, the brothers would make the daily trek to Meadfoot to join Russell on the beach. In turn, the young American invited them to spend the following summer in New England. He couldn't understand why the Owens were unable to accept.

As the summer stretched ahead, Wilfred and Harold were left to their own devices and diversions – among them one of the most splendid spectacles at sea that the country had ever seen, as the new king-emperor arrived to review his fleet. But the focus that day subtly shifted, from the monarch to another figure.

Claude Grahame-White was Britain's most celebrated and handsome aviator, a former car salesman whose aeronautical outfit consisted of a three-piece knickerbocker tweed suit, cap and tie, sportily set off with a cigarette dangling nonchalantly from the corner of his mouth. His aeroplane looked more like a mechanical bird than a flying machine: a fragile thing of wood struts, doped linen and animal glue, yellow-skinned and thin-boned, held together by faith as much as by technology. Its appearance, one thousand feet over Torbay, was akin to the sighting of a UFO; the first time most of the crowd – which had been, up until that moment, an Edwardian assembly with straw boaters and picture hats and long skirts, craning their necks to watch the plane swoop over the bay – had ever seen anything other than a seagull in the air. Suddenly, they were modern. They were looking at the future. A man, flying.

The Farman III biplane took off from the beach-side field in front of Torre Abbey, flying over its gothic arches and the Georgian mansion where Nelson had once dined. Biplanes generally flew at dusk, when the day's winds had abated. Grahame-White was borne up on a thermal, rising on the summer heat that had kept children paddling in the water all day. The aircraft glowed as sparks from its engine lit its linen wings like a Chinese lantern. Suffused with the colours of twilight, the scene might have been painted by John Singer Sargent.

Yet the pilot's intent was anything but glamorous. The next day, on his second launch, he flew in the cold light of dawn. It was a foggy morning, and it seemed the review would have to be called off. But then the mist lifted to reveal eight columns of ships of the Royal Navy, sea-going enforcers of the greatest empire the world had ever seen. It was a unique assembly, an absolute demonstration of imperial power and industrial mastery of the sea: one hundred million pounds' worth of weaponry. These 'engines of war' were an index of dominion, sweeping over heroes and victories and kings and colonies and out to the stars themselves: *St Vincent, Collingwood, Lord Nelson* and *Temeraire*; *Edward VII, Hibernia, Africa* and *New Zealand*; *Dominion, Commonwealth, Hindustan* and *Britannia*; *Vanguard, Superb, Bellerophon* and *Jupiter*. Not even the underwater world went unpatrolled. Six submarines – led by the leviathanic D1, built in utmost secrecy in Scotland and showing 'like a salmon among minnows' in the mechanical shoal – rounded Berry Head to the surprise of its dolphins and porpoises, and slid into place in the bay.

George V's ascent to the throne, the reassurance of the same, seemed a bulwark against the changes of the new century. But that summer, as Wilfred and Harold swam and children played on the beach, Britain was filled with war games. Territorial troops were training on Salisbury Plain, and in Kent soldiers practised with machine guns on night operations. In Wales, the Royal Welsh Fusiliers were faced with the prospect that a revolt in Ireland had been followed by a landing of forces north of Aberystwyth. And in a vast exercise carried out on the Firth of Forth, hundreds of troops, including the Cyclist Battalion of the Royal Scots, fought an imagined invasion landing at Dunbar, intent on marching on Edinburgh. Every shore of the kingdom seemed threatened; the country was preparing for attack from every direction, even from under the sea.

The militarised new monarch – 'it must not be forgotten ... that the King himself has handled a torpedo-boat, voyaged in a submarine, and commanded various classes of vessels from a gunboat to a big cruiser' – watched from the armipotent battleship *Dreadnought*, the incarnation of imperial might (for all that a few months before, Virginia Woolf, along with five of her Bloomsbury male friends, two of whom were lovers, had conducted their own stunt, boarding the ship in Weymouth wearing costumes and blacked-up faces, pretending to be a royal delegation from Abyssinia).

As the monarch peered up through his telescope, the biplane passed directly overhead. To *The Times*, it was an uplifting sight; Grahame-White was exhibiting a new kind of art. 'He made a particularly pretty flight, rising easily to about 1,500ft ... then gliding round to a lower level to give the King an opportunity of seeing him more closely as he flew over the Dreadnought.'

The reporter was caught up in wonder like the rest of the crowd; this was aerial choreography to rival any Russian ballet. 'All his gyrations and movements were made with that ease and smoothness for which he is noted, and his swoop to the landing-place was particularly workmanlike and beautiful.' And as he rose over the trees under which their carriages were gathered, the crowd looked up as one in wonder. Their horses stood straight ahead, eyes blinded by leather blinkers.

At five o'clock that afternoon the aviator made his third flight. The fog was thickening, putting an end to the naval exercises, and the ships were coming in. As he soared over *Dreadnought* once more, Grahame-White dipped his wings and saluted his monarch.

The king waved his telescope-sceptre from the deck of the world's most mighty ship. He did not realise what had changed in that instant.

The pilot had proved his point. None of the navy's guns, able to quell rebellion and enforce diplomacy in every ocean, could swivel upwards and – had Grahame-White been an enemy – take him down. This Icarus in a tweed suit, borne aloft by linen wings and perched on what was little better than a kite with a car engine, was an augury of the time when such frail contraptions would become terrible firebirds, dealing not delight but death from above. As any classically-educated observer in the crowd might have noted, the word augury referred to the Roman belief in determining the future from a bird in flight, and shared the same root as avian and aviator.

That night, Wilfred and Harold stayed out on the cliffs until midnight. Looking down on the promenade strung with lights and dotted with palm-like cordylines, they pretended Torquay was some city in the tropics, shimmering under the dark-blue sky. Out at sea, the fleet was also lit up, and every now and again a ship would send up a rocket, its trail reflected in the black waters of the bay like a falling star.

The following spring, Wilfred returned to Torquay, alone. He missed Russell, who was now at Harvard, and in solitary moments the young poet sat on the rocks at Meadfoot, reading a French book he'd bought, filled to overwhelming with the sea. As a would-be university scholar himself, Wilfred translated a passage from Alphonse Daudet for his mother's benefit: 'You know, don't you, this lovely intoxication of the soul? You are not thinking, you are not dreaming either. All your being escapes you, flies off, is scattered. You are the plunging wave, the dust of foam which floats in the sun between two waves ... everything but yourself.'

'Well, I was reading the book at Meadfoot the other afternoon,' said Wilfred, as he broke off from his reverie, 'when who should I see but the boy-youth whom we met with Russell Tarr, and who played croquet with us on the lawn, as Harold will ...'

But the rest of the letter is missing, as are Wilfred's other letters about Russell, censored by his brother in order to preserve his brother's reputation – at least, as he saw it. Harold blocked Wilfred's words with black ink; often he cut out paragraphs or removed entire pages. What was left made a mystery of what might have been banal, or perhaps not so. That not-knowing, the withdrawal of permission, is physical, like someone holding their hand in front of your face. He was doing to Owen's letters what the sea had done to Shelley's notebook; only instead of the water washing out the words, thick squid-black ink covered their traces. The result was quite contrary to Harold's intent. He only made the ordinary more extraordinary.

It wasn't the job Wilfred had imagined back in Torquay, but it would do for now. He was hurtling through the country lanes of Berkshire on his magical Sturmey-Archer three-speed bike, attending to his duties as assistant to the vicar of Dunsden. Cycling, like swimming, was a release; it took him into the natural world. He was a keen bird-watcher, and admired swallows so much that his family called them Wilfred's swallows. Such fast birds fascinated him, built only for the air, hardly ever touching the ground, yet travelling from one hemisphere to the other. He had recently written a Keatsian ode to the swift, 'Airily sweeping and swinging, | Quivering unstable', and as he cycled below, their wings carved their own curves into the sky: 'If my soul flew with thy assurance, | What Fields, what skies to scour! what Seas to brave!'

Despite its sententious reverend and the deadening sound of the vicarage clock ticking away the afternoon, Wilfred liked his new post. Dunsden sounded dull, but it lay close to the Thames; he was excited to learn that Shelley had lived in a cottage by the river in Marlow, where he'd erected an altar to Pan in the woods.

A photograph of Owen in the vicarage garden shows a typical teenager. He sits with his legs crossed and his body at an angle; a short, neat young man, just five foot five.

Perhaps it's Sunday, the dullest, freest day of the week. He's wearing a checked Norfolk jacket, flannel trousers, a Homburg hat by his side. His hair is side-parted. There's a critical look on his unformed face as he reads aloud from a book held in one hand, the other curled under his chin. He may be serious; he may not. I imagine him arguing passionately from under that floppy hair and furrowed brow, all taut with complexities, even as you can hear him complaining about being told to tidy up his room when he has more important work of his own to do.

He had just returned from the Lake District, where he'd spent a week at an open-air Evangelical gathering, his generation's version of a music festival. He was devout despite his doubts, and in his letters – he would write five hundred to his mother during his lifetime – his voice comes alive, funny and questioning and energetic; so many words, as if he were cramming them all in. At this field camp, with its endless hymns and sermons, he'd slept in a smelly tent and sung along with all the other young men, finding only one, a mining lad, interesting. When he took his shirt off, the boy's back showed the scars of his subterranean work; to Wilfred, he seemed somehow angelic, 'tho' pricked with piercing pain'.

Escaping the camp's holy orders, Wilfred swam in the nearby lake, and cycled forty miles in the pouring rain to Coniston to see Ruskin's house, Brantwood, set over the slate-grey water. Ever the fan, Wilfred called on Ruskin's secretary and biographer W.G. Collingwood, who lived nearby. I don't know if he swam at Coniston too; I did, and found it fearful with the memory of Ruskin's madness, imagining the critic-prophet watching the storm clouds of the nineteenth century from his gothic turret overlooking the lake. That night Wilfred rode back through the storm, 'drunk on Ruskin' and utopias and possibilities.

Now he was back in his routine, on his bike, making pastoral visits to elderly parishioners, delivering not food but comforting words from the sermons of Charles Spurgeon, 'about human lives being as frail as flowers, as fleeting as meteors'. Of course he wanted to be a meteor too, 'fast, eccentric, lone, | And lawless'. Astronomy fascinated him: that April he had witnessed the greatest solar eclipse for fifty years (for which Virginia Woolf had travelled to Yorkshire), and in an earlier letter he had sketched a new comet he'd seen in the night sky.

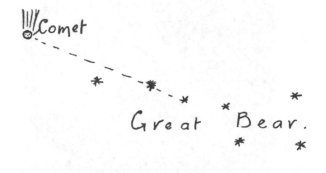

Wilfred rode on with his head full of all these things: with Keats and Shelley and Ruskin and shooting stars and comets and his own poems as he hurtled through the countryside, when suddenly his bike skidded and he slid violently to the ground.

The incident lasted barely ten seconds. Dazed and in shock, he managed to ride back to the vicarage, where the housekeeper took one look at his white face and the deep cuts on his hand and sent for the doctor. Wilfred started to lose consciousness.

'Sudden twilight seemed to fall upon the world, an horror of great darkness closed around me.' He heard the blood roaring in his ears, and 'strange noises and a sensation of swimming under water'. Although he did not faint, he broke out in a cold sweat and, somewhat deliriously, found himself 'gasping at a window, without quite knowing how I got there'. He went to look in the mirror, in which he admired his 'really beautiful and romantic pallor', and claimed, 'I had a presentiment that something of note would happen shortly.' Wilfred was all too sensitive to omens; around this time he made a note on one of his loose pages: 'why have so many poets courted death?' He was in love with death, as boys are. It was the biggest adventure.

A few days later, Wilfred resumed work on 'The Little Mermaid', now into its second exercise book. He sent his heroine diving to the sea floor to find 'a marble statue, – some boy-king's, | Or youthful hero's', whose cold face she kissed. The drowned image, out of Coleridge and Shakespeare, foresees her fatal meeting with her prince, for whom she forsakes the sea, her elegant tail bisecting into two painful legs, as if she were walking on knives, like Byron. The poem ends as she helplessly pursues the prince's ship into a storm, 'All ears | Hark to a grumbling in the heart of the seas.'

On the Friday after Harvest Festival, Wilfred got back on his bike and rode to his cousins' house for a family reading of The Tempest. The play was a set text on his correspondence course in his attempt to get to university, and he had recently seen it performed in London. The elemental drama touched him with its strangeness; it spoke to the distance he felt between his art and his nature; between the future and the past, between what people expected and what he wanted to be. Later that year he went to see it in London for a second time, and from the train on the way back he saw a biplane 'high in the Western sky', so high he could not hear it. The machine, from Grahame-White's new aerodrome at

Hendon, was all the more impressive for its silence against the setting sun, held in the windless air.

In April 1913, worn down by his duties at Dunsden, Wilfred returned to Torquay to recuperate from congested lungs; the sea air would clear them, so he and his mother, who feared the first signs of consumption, hoped. He found the resort a changed place. His uncle had died, and his aunt had moved to a smaller house. Wilfred revisited Teignmouth, and both there and at Babbacombe he observed that the swelling afternoon tide seemed to bring on the rain, which his scientific mind sought to explain: 'the rise of such a vast surface of water ... compresses the Air above to a greater density'. But as the clouds lifted, so did his depression about the dead end his life had reached, having failed to get into university. The 'verray' blue sea worked its magic – although, as he told his mother, there were three battleships in the bay, closer to shore than he had ever seen. 'How melancholy-happy I was,' he wrote, quoting Keats, his consumptive hero, 'where the wide sea did weave.'

Wilfred needed to escape, and the opportunity arrived: the offer of a job as a teacher in Bordeaux. As he crossed the Channel, the world suddenly opened up. He was no longer a boy from a backwater. He was a new European man. He grew a moustache and wore his hair fashionably slicked back and centre-parted, and acquired a tan which would stay with him for the rest of his life. He even fantasised about rebuilding his body with international parts, like a mail-order version of Mary Shelley's Creature. He chose eyes from France, mouth from Italy, hair from Greece, chest from Sweden, legs ('badly needed') from France also, and shoulders from America, 'of course'. At twenty-one, he placed great store in his use of the word 'Beautiful'.

The reality was somewhat different. Wilfred suffered hypochondriacally, detailing to his mother every symptom, living on chocolate and raw eggs and bemoaning the fact that his old green suit was wearing out. (He favoured green: it was the aesthete's colour, as sported by Wilde and his acolytes. Later, playing a dandy in an amateur drama at Craiglockhart, he'd ask his mother to send up his green suit, green shirts and green glass cufflinks for his ensemble.)

Wilfred had accepted a new post, as tutor to the Léger family: Charles Léger was the founder of the experimental Théâtre des Poètes, and impressed Wilfred because as a boy he'd met Mrs Browning, who had shown him particular kindness; Madame Léger, his young wife, Wilfred's student, ran an interior-design company. They were staying at their summer house in the Pyrenees, a villa in the Belle Époque style on the outskirts of Bagnères. The town's name meant baths, famed since Roman times for its healing waters. Lourdes was only a few miles away, with its own miraculous spring.

Wilfred was introduced to a southern life. Dressed in a consciously cosmopolitan manner in straw boater, bow tie and smart ankle boots at an open-air reading, he sat next to Madame Léger – who rather fancied him, he knew – as they listened to the poet Laurent Tailhade, who rather fancied him too. A decadent anarchist and opium addict, and friend of Verlaine and Mallarmé, Tailhade had declared a terrorist bomb in the French parliament to be a 'beautiful gesture', much as Karlheinz Stockhausen would call the attacks of 11 September 2001 a work of art, or as Turner had hired a boat to paint the Palace of Westminster while it burned to the ground in 1834. Tailhade still supported the anarchists even after he was injured by one of their bombs and lost the sight of one eye. Under his radical, pacifist influence Wilfred's ideas changed. He too would find art in disaster.

From the Légers' villa Wilfred would walk into the hills, where he discovered a mountain stream in which to swim. His hosts advised against it: too cold, not deep enough, people might be about. Wilfred dismissed these prissy reservations. Having found 'an enchanting stretch of water in an alder-glade, I was not long in "getting in". So now I go down every day, and I know, when I vaunt my coolness and freshness, the others would be green with envy, if they were not so infernally red with heat.' He might have been Shelley dipping naked in a neoclassical landscape. But this was August 1914, and in Futurist-inflected pararhyme, Owen recorded that last summer more surely than if he'd photographed it.

Boys
　　　Breaking the surface of the ebony pond.
Flashes
　　　Of swimmers carving through the sparkling cold.
Fleshes
　　　Gleaming wetness to the morning gold.

He knew what he was doing. He was becoming a modern artist, physically expressed in his body. Even here, in the soft light of southern France, the seismic shocks of the war that would transform him were felt as disturbances in the air. Women cried in the streets, and Wilfred looked to the mountains of Spain in the distance, wondering if he ought to take refuge there. The wounded were arriving from the front; he saw German prisoners of war kept in cages. 'I like to think that this is the last War of the World!' he told his younger brother Colin, as though sharing some science-fiction comic.

Charted in his letters, the slow, inevitable speed of Owen's assumption into war is frightening and obvious, as if he were cycling into it. I look at his bookended dates and wish I could have written a different biography for him. Perhaps if he hadn't written it all down, or if I hadn't read it, it might not have happened. I might have seen him in the nineteen-seventies, an old man with a greying moustache on Meadfoot beach, looking out to sea, looking back into a past that never happened.

Giving up teaching, Wilfred returned to Bordeaux and became a perfume salesman – a somewhat decadent occupation in war-time, although Stephen Tennant would have approved – then returned to London, crossing a sea sown with mines, patrolled below by U-boats and threatened overhead by Zeppelins.

Yet even this close to the point of no return, he was torn between art and war. He walked through the East End and down to the river at Limehouse, a tidal place with its own secrets. He saw a 'godlike youth … with blood-red lips' whom he dared not approach, and who merged with a shadow he met on Shadwell Stair, its steps leading into the Pool of London. This boy's eyes reflected 'moons and lamps in the full Thames', and this water – another kind of danger zone – mirrored Owen's desires, 'Like the sleepy tide upon the sands | To feel and follow a man's delight'.

But such Wildean scenes had been banished by a strident new voice. Rupert Brooke's sonnet saw the fighting as an antidote to ennui: 'To turn, as swimmers into cleanness leaping, | Glad from a world grown old and cold and weary'. To Wilfred, a stark choice presented itself. 'I have made soundings in deep waters,' he told his mother, 'and I have looked out from many observation-towers: and I found the deep waters terrible, and nearly lost my breath there.'

He took a headlong dive. Once his decision had been made, the transition was abrupt, and shocking. On 21 October 1915 Owen was sworn in at the headquarters of the Artists Rifles in Bloomsbury; given that he felt he was enlisting to defend poetry, his choice of recruitment station, in an area of modernist dissent inhabited by Woolf and her war-resisting peers, is ironic. Within weeks his body was subordinated and subsumed into the military machine at camps in Essex and Aldershot, far from Lake District evangelicals, and even further from elegant French villas and the proffered embraces of older women and one-eyed anarchist poets.

A photograph taken around this time, only recently discovered, shows Owen in an ill-fitting greatcoat, handed to him straight from the stores. Its thick, rough wool – more like a blanket – is pulled in at the waist by a webbing belt. It swamps his little body, hanging off his shoulders. But what surprises most is the smile on his face. If it weren't for the rather tentative moustache, he'd look like a ten-year-old who'd just scored a goal. He grins from ear to ear, showing off his white teeth and his dimples. He looks like a boy you knew at school. He simply seems pleased to be a soldier.

By the summer of 1916, the transformation was complete. His fringe, which once fell carelessly over his broad forehead, was cropped ('Note I wear my hair ½ in long now'), so brutally that his brother would seek to censor these images too. Harold was amazed when he called unexpectedly on Wilfred in his barracks that September to find him 'clad in khaki slacks and nothing else', his bare torso impressively developed, toughened by training during which, like other officers, he had grown an inch in height.

Owen the suburban boy had become a sophisticated officer, with offhand manners to match. He rose from his camp bed and dressed himself, assuming an elegant persona ordered from smart London shops. His Pope & Bradley tunic, waisted and epauletted, emphasised his new body; his torso was restrained by the Sam Browne belt slung across his chest, the leather lovingly buffed with a velvet shoe pad. His neatly parted, cropped hair was slick with brilliantine and groomed, like his clothes, with ivory-backed brushes from Swaine Adeney. The silkily knotted tie and collar-pin, the gleaming boots and riding crop – these were all part of a modern poet's armour in time of war: close-fitting, tailored, belted, buttoned and buckled. He was laced and strapped and tied and bound by the state, right down to his putteed legs. It was as if Wilde had been sent to war rather than Oxford; the aesthete's green had turned into military khaki, as though already muddied by the trenches. His fellow officers were 'desperate nuts in dress: as befits our calling, and immemorial custom of all gallants'; their favourite 'Flash' was to sport a violet handkerchief trailing from their khaki pockets. Wilfred felt that was a bit too flash – and a dishonour to that noble colour, purple. Nevertheless, he doused his handkerchief in scent to stave off the stink of sick horses, much as a Regency dandy had recourse to his vinegrette in the odiferous streets of Piccadilly.

Photographed by his uncle that summer, his face is startlingly modern. He could be anything, but this is what he has decided to be, for this particular moment. He is boyishly tanned, gazing under his peaked cap through grey-brown eyes, faintly smiling under his 'soldierly moustache', knowing, as we all do at that age, that deep inside he was not really a part of this at all; that this was just another skin which would eventually be shed. 'Outwardly I will conform,' he told Harold, 'my inward force will be the greater for it.' Finally his name was in print, but not in the way he'd imagined –

W.E.S. Owen, CDT

– reduced to tiny initials in the gazette pages of The Times, signifying nothing more glorious than an assimilation into the system, along with column inches of other officers. Only when you see him next to those fellow officers – of the Fifth (Reserve) Battalion, under the pines of their Surrey campsite – do you realise how small he is, and how different, for all his efforts to look the same. He might be a particularly elegant Boy Scout. Yet he was now in charge of other men, younger and older than him. Power and duty became him. His instincts lay with the other ranks; his snobbish side allied with the officers: 'I am marooned on a Crag of superiority in an ocean of soldiers.' As a second lieutenant out training his platoon, he could break off from his duties on a hot summer's afternoon by the side of a Hampshire lake – the order having been given for bathing to be allowed – and wander to a secluded cove where he found 'a solitary, mysterious kind of boy', the son of a Portuguese aristocrat, to swim with. He had a knack of finding such temporary companions. We forget, perhaps, what a charming man he was; or how sensually charged a war can be.

And Owen was a remarkably *good* soldier – good with his men, and an expert shot. Having proved that he could serve, that he had a place, he was determined to become an even more ambitious part of the war effort, in the Royal Flying Corps. 'By Hermes I shall fly,' he told his mother, turning himself into a young god. 'I will yet swoop over Wrekin with the strength of a thousand Eagles ... the pinion of Hermes, who is called Mercury, upon my cap. Then I will publish my ode on the Swift. If I fall, I shall fall mightily. I shall be with Perseus and Icarus, whom I loved; and not with Fritz, whom I do not hate. To battle with the Super-Zeppelin, when he comes, this would be chivalry more than Arthur dreamed of . . .' Earlier, he had seen his first Zeppelin in the skies over his Essex camp, its whalish body picked out in the searchlights. 'The beast looked frightened somehow,' he wrote, as he watched it nose about as if lost, only to vanish into a cloud which, he thought, was of its own making.

Wilfred was thinking of his favourite painting, *Lament for Icarus*, by Herbert James Draper, which was on display in the Tate Gallery, on the banks of the Thames. Draper liked to depict nude sirens decorously draped with seaweed in an imaginary sea; his Icarus, painted in 1898, sprawls on a rock in blue-green water, as if dragged out of Brueghel's sea, or the Thames at Limehouse. The youth's semi-naked body has been tanned by his close approach to the sun; he lies spreadeagled and folded in his huge but defunct wings, which Draper modelled on those of a bird of paradise.

But a painting in the Tate, no matter how heroic, was no substitute for the real thing: 'Betimes I have a horrible great craving to behold the sea.' Wilfred's wish was granted that autumn, when he was sent to Southport for training – only to discover the tide there ran out so far over the wide flat sands – where half a century before Herman Melville had met Nathaniel Hawthorne and confessed his thoughts of self-destruction – that it was the 'most unsatisfactory sea-side place in Europe'. Wilfred was under military manners but he could have still been a boy, writing home after leaving Torquay, 'how I miss my morning bathe'. Within weeks he was on his way back to France; not as a young man in a dapper suit in pursuit of sensation, nor even a newly-fledged hero, but as a uniformed officer charged with the absolute implementation of violence. Those few miles of water between his island home and the embattled continent represented a new gulf. He already looked back to another life, 'the days when my stars were bright from their creation by Pope & Bradley'. Inexorably, he was ordered to war, to experience its pity, and its pitilessness.

As he got nearer the front, Owen was physically assailed by the noise. It both repelled and drew him on. His duty was to minister to his men, much as he had ministered to the parish of Dunsden, 'and this very day I knelt down with a candle and watched each man perform his anointment with Whale Oil'. Like a priest at the Mass of the Last Supper, he supervised them as they prepared for the flooded trenches by rubbing their bare feet with rendered-down blubber.

For these men, their future condition would be amphibian; this was a fearful morning bathe. Owen and his company were sent into the worst conditions on the Western Front, where the land had become 'an octopus of sucking clay', navigable only by aquatic duckboards. Bomb craters were bottomless lakes in which men drowned; others only managed to extract themselves by leaving their equipment and even their clothes behind – a weird, Stanley Spencer scene of resurrection, white bodies reborn out of primordial mud. These warriors were dressed ready to fish for some evil prey: Owen was 'transformed now, wearing a steel helmet, buff jerkin of leather, rubber-waders up to the hips & gauntlets. But for the rifle, we are exactly like Cromwellian Troopers,' lacking only the lobster-tailed helmets of the New Model Army. He could hardly breathe under his tin hat; when he took it off at night, he took in lungfuls of air, as if he'd been bolted into a brass diving helmet. He seemed to be walking on the sea bed. 'In 2½ miles of trench which I waded yesterday there was not one inch of dry ground. There is a mean depth of 2 feet of water.'

It was a nightmarish version of those seaside days. He was weighed down; sinking, not swimming. Water had become protean and terrible, as it had for Elizabeth Barrett Browning and her memory of her drowned brother, evoked in *Aurora Leigh*: 'When something floats up suddenly, out there, | Turns over ... a dead face, known once alive –'. Arriving in the bombed-out village of Bouchoir, Wilfred found a copy of Mrs Browning's collected poems, and in the battered book he underlined a passage from *Aurora Leigh*: 'See the earth, | The body of our body, the green earth | Indubitably human, like this flesh.' She was following him into the mire, picking her way through the devastation in her dark crinoline gown. Her words accompanied him as his mind wandered from the primordial mud to his internal ear, 'having listened so long to her low, sighing voice (which <u>can</u> be <u>heard</u> often through the page,) and having seen her hair, not in a museum case, but palpably in visions . . .'

But this was no place for poets. This was nature suborned, perverted, and destroyed. Brooke's clean swimmers wallowed in flooded trenches and their own dung. The countryside looked more like the blackened mud and debris left on the beach at low tide, with explosive shells instead of their marine equivalents and the buried dead constantly disinterred by each new wave of fighting. It was an inundated deathscape no silent film could record; although perhaps Shelley had foreseen this apocalypse, in which machines spewed Promethean fire and raptors pecked at heroes' bones.

At one point Owen's platoon took shelter in an empty German dugout, accessed by a tunnel which descended as deep as the height of a house. They remained for almost two days in this earthy chamber, like some prehistoric barrow, up to their knees in water which was rising ever further, while the enemy fired directly at them.

The men shook, spewed and shat themselves in fear. At one point the sentry Owen had posted at the top of the tunnel was hit, and came tumbling down the steps, 'sploshing in the flood, deluging muck'. He was not dead, but he could not see.

'O sir – my eyes, – I'm blind, – I'm blind, – I'm blind!'

Owen held a candle to his face, telling him that if he could see 'the least blurred light', he'd be all right. But he sobbed his reply, 'I can't.' His 'eyeballs, huge-bulged like squids'', would haunt Owen for the rest of his life.

Everything was overcome. In his own desperation, Owen even considered allowing himself to succumb: 'I nearly broke down and let myself drown in the water that was now slowly rising over my knees.' Water had become a fetid medium in which war bred: as though Turner had reprised his *Sunrise with Sea Monsters* out of the Flanders mire, lashed to a lumbering tank – a land ship – while other armoured leviathans surfed over shell holes and sunken lanes like his scaly beasts, watched all the while by the helmeted enemy through submarine periscopes.

All the devils really were here. 'Hideous landscapes, vile noises, foul language and nothing but foul, even from one's own mouth (for all are devil ridden), everything unnatural, broken, blasted,' Owen told his mother, invoking a seventh hell, scattered with the distorted bodies of the dead, 'the most execrable sights on earth'.

Claude Grahame-White's salesman's prophecy had become a real war of the world, an alien invasion of machines that aped animals. Wasp-like aircraft buzzed overhead while whale-like Zeppelins, ominous five-hundred-foot-long, skin-covered shapes sailing over the German Ocean, dropped bombs on English re-sorts and even hovered over the shore where I swim. Britain was no longer safe as an island. The sea that surrounded it was now a conduit of disaster rather than trade or pleasure. It boiled and bled from below, as Virginia Woolf would write in *To the Lighthouse*. As the submarine peril increased, fear drove some to seek super-natural aid: the cost of a talismanic caul increased from one and six to three guineas for those who wished to protect their loved ones from drowning.

Meanwhile whales were mistaken for U-boats and blown up, or used as target practice, while their blubber was processed into nitroglycerine. The demand was sated by the slaughter of eighty thousand cetaceans, their bodies processed in remote whaling stations by other hip-booted men. Animals became by-products and victims of a great war in which as many horses died as men. It was a war for natural resources, the first war of the Anthropocene, an augury of extinction. In 1917 Siegfried Sassoon was appalled to discover that the real aim of the conflict was 'essentially acquisitive, what we were fighting for was the Mesopotamian Oil Wells'. The machine was all-consuming, killing more whales and sinking more wells into oceans of oil, producing the raw material for more war. It was an inequitable, international exchange rate. In her diary, Woolf recorded that to kill one German at the Somme in 1916 cost Britain one thousand pounds; a year later, the price had risen to three thousand.

Sunk deep in those trenches, men donned aqualung-like masks against the waves of gas in 'an ecstasy of fumbling'. One soldier unable to do so in time was watched by Owen: 'Dim, through the misty panes and thick green light, | As under a green sea, I saw him drowning.' The poet's nerves, already raw, were brought to a state of exposure in those nine days at the front, each day worse than the last. As he slept on a railway embankment, a large shell landed barely two yards from his head, blowing him into the air, his body briefly leaving the earth. It was another transformation. A rebirth, of sorts.

At first it seemed he had only slight concussion; but a few days later he was observed, as his army file records, 'to be shaky and tremulous, and his conduct and manner were peculiar, and his memory was confused'. He was suffering from shell-shock, brought on not by the explosion, but by the fact that he had lain in a pit for two days, sheltering under a sheet of corrugated iron next to his fellow officer, Second Lieutenant Gaukroger, who had been killed a week before and who now lay around him, in pieces.

Slowly, he was moved back to the sea. Étretat, a resort twenty miles north of Le Havre, was famous for its chalk cliffs, arches and stacks, much painted by the Impressionists. It too had become an anteroom of war, filled with khaki uniforms and materiel going one way, and doctors and nurses dealing with the results on the return journey. But it also represented the clear blue sea he loved. He was kept in a marquee on the lawn of a hospital run by the

Americans – a reminder of his swimming friend Russell, himself now serving in the war, having found wings of his own in the intelligence department of the US Army, preparing maps for the 29th Engineers. It was as if Wilfred had finally been able to take up Tarr's offer; like Woolf, he conjured up a continent he would never see.

'I seem to be in America,' he told his mother, 'on this delicious Norman Coast,' adding a little drawing of the Stars and Stripes to his picture postcard of Étretat's cliffs. Warmed by the sunlight filtered through cream-coloured canvas, he could believe he had gone to heaven. 'This is the kind of Paradise I am in at the present,' he told Colin, his youngest brother. 'The doctors, orderlies and sisters are all Americans, straight from N.York! I may get permission to go boating & even to bathe.' He couldn't quite comprehend that he would soon be back in England – 'I shall believe it as soon as I find myself within swimming distance of the Suffolk Coast,' he declared, as if he might flout all authority and swim back home. Instead, he contented himself with the prospect of the French beach: 'If I go bathing this afternoon it will be to practise swimming in Channel waters.' He was only a few miles from the shore where Mrs Browning had come to dip her body in salt water.

A few days later, Wilfred was taken back to England by converted West Indian liner – with the luxury of a cabin to himself – not to the Suffolk coast, but up the Solent to Netley's vast military hospital. 'We are on Southampton Water, pleasantly placed,' he told his mother, 'but not so lovely a coast as Etretat.' The sea overcame words and what he had witnessed. 'Nothing to write about now. I am in too receptive a mood to speak at all about the other side the seamy side of the *Manche*. I just wander about absorbing Hampshire.' He felt a sense of disconnection, in his abrupt transition from the front and its deafening death cult to a semi-rural site with its main building an endless brick terrace topped by a verdigris dome, and behind it, rows of wooden huts. (One newspaper reported, 'We are reverting to primitive ways. Like disciples of Thoreau we have gone forth and built huts in the woods and by the waters.') He wasn't impressed with my hometown. 'This place is very boring,' he decided, 'and I cannot quite believe myself back on England in this unknown region.' He added a sardonic caption to his postcard of the huge hospital, calling it a 'Bungalow'.

In this sprawling medicropolis-cum-military resort, a half-way house between the martial and the civilian world, soldiers exchanged their khaki for pale-blue uniforms known as hospital pyjamas. They were loose enough to evoke holiday clothes, per-haps; but they were accessorised by blood-red ties. As an officer, Owen escaped this indignity; he merely wore a blue armband to signify his changed status. He was in limbo, awaiting his assess-ment as a mentally- rather than physically-wounded man. Half of all shell-shock casualties from the front were cleared through Netley; the fields teemed with traumatised men, a war-damaged crop. Wilfred may not have been able to believe himself in this place as he walked its beach, but I could. Swimming off that same shore, I look over and expect to see him trudging through the shingle, talking to himself.

After a week at Netley, and a stopover in London – where he boasted of writing a letter from a Piccadilly teashop under an opium den, perhaps remembering his decadent friend Tailhade – Owen arrived at Craiglockhart, a down-at-heel hydropathic establishment outside Edinburgh, complete with swimming pool and Turkish baths. It was now occupied by shell-shocked officers shipped there from flooded trenches; some wore nothing but borrowed bathing costumes from the pool. The house magazine, which Owen would edit, was called the *Hydra* after the mythical many-headed water-snake; but equally it might have evoked the snarling monsters that haunted the men's dreams.

Military medicine fed on its own victims. But encouraged by his enlightened doctor, Captain Arthur Brock – who had studied in Vienna and who used as a teaching aid a print of a classical sculp-ture of Hercules and Antaeus wrestling – Owen tried to readjust, 'chiefly by swimming in the Public Baths really religiously, for it never fails to give me a Greek feeling of energy and elemental life'.

An Edinburgh librarian who met him around this time described Owen's 'comeliness': his features fluid and sharp, his compact body like that of a boy, with the muscles of a man; a century later I would meet Owen's nephew – he had that same short body, that same broad forehead. Yet Wilfred's physical recovery was not reflected in his face. It had tightened; the slight smile and dark eyes now held a haunted look. It is the same change I see in photographs of my own grandfather: a handsome young soldier in 1914; a haggard man with a lined face just five years later.

I cannot imagine what these men saw. Owen dreamed of his sentry falling back blinded, and of motor accidents, and of a man who received a shrapnel ball 'just where the wet skin glistened when he swam', a round red hole 'like a full-opened sea-anemone'; the wound became infected and he died on the way home, buried at sea along 'with the anemones off Dover'. The delayed impact of what he had witnessed infected his semi-civilian life. Others who met him then claimed that Owen carried photographs of dead and wounded men in his pocket, a gallery of horror which he would produce as proof of the true effects of war; but in reality, the poetry he was about to write would prove far more effective. By day he wandered the streets of Edinburgh; at night he returned to the trenches. If this city had inspired Jekyll and Hyde, then Stevenson's story was replayed in the personae of Owen's fellow officers. At any point they could switch. One moment his friend, a young officer named Mayes, was perfectly normal. The next, he appeared at Wilfred's door with staring eyes, mouthing words he was unable to utter, making strange gestures with his hands.

Siegfried Sassoon was wearing a purple 'dressing suit' on which the sun was shining brilliantly when Wilfred knocked at his bedroom door and asked him to sign his latest book. Sassoon, a hero to Bloomsbury for his rejection of the war's aims, was of another class entirely: he would refer to Wilfred as 'little Owen', and thought him 'perceptibly provincial', but also 'a very loveable creature'.

For his part, Wilfred fell in love with Sassoon, innocently, knowingly. Being blown into the air had turned his life around; it was as if he was still falling back to earth. 'I was always a mad comet,' he told Siegfried, 'a dark star in the orbit where you will blaze.'

Now everything changed again. He was introduced to the in-crowd. In Edinburgh, Wilfred was invited into an artist's house where the floor was black and the walls were white, and his hostess wore bright clothes and had her hair cut short. In London he became a house guest of Robbie Ross, Wilde's lover and literary executor, who had painted his rooms on Half Moon Street gold as a protest against the war, and where every night Turkish delight, brandy and cigarettes were laid out for any friends who might call. These were no unknowing acts on Owen's part. When he arrived at Robert Graves's wedding at St James's, Piccadilly with Ross – all but on the older man's arm, led in like a faun found in the woods – it was a statement of allegiance. At the party afterwards he met Charles Scott Moncrieff, poet and translator of Proust, who would fall in love with him; both Scott Moncrieff and Ross were about to become entangled in the Pemberton Billing trial, a scandal of conspiracy and prejudice stirred up (not least by a now-embittered Bosie) against Wildean decadence. Wilfred wrote home that he was now 'one of the ones'. Wilde's heirs were dangerous people. Owen knew their power; he knew his own. As his biographer Dominic Hibberd wrote, his assault 'on the civilian conscience' was a wartime version of the 'Decadent urge to shock'.

But there was so little time left. In one month, October 1917, he wrote 'Anthem for Doomed Youth', 'Dulce et Decorum Est' and 'Disabled'. He had barely a year in which to compose the other poems that would make him famous. He told Harold, when they met as soldier and sailor, that his work and the war were 'running a ghastly race' – adding, 'I am getting very tired and short of breath, but I don't think the War is.'

At the end of October, Owen was judged fit for duty. At first it seemed that he would be kept on the home front. He rejoined his regiment at Scarborough, another resort familiar from his childhood; the beach where he and his family rode horses on the sands was now lined with barbed wire. Scarborough had been shelled by the German fleet in 1914; only two months before Wilfred arrived, it was attacked by an enemy submarine. Some claimed that a German officer had landed from a U-boat and visited one of the town's pubs.

Wilfred was quartered in a smart hotel with a turret, on the cliff overlooking the North Sea. In between duties as a 'major domo' – effectively, a housekeeper to his fellow officers – he went shopping for antiques for his future home, which he had designed even as he lay amid the mud of the trenches, sketching a seaside bungalow, complete with colour scheme and carpets, to which he would escape, like Wilde in his beach hut, once the war was over. After drinking in the town's oyster bar, Wilfred retreated to his turret from where he could look down on the waves and, wrapping himself in his dressing gown with his feet in purple slippers, became a poet again.

He wrote poems filled with barefoot 'little gods' and 'youthful mariners', their thighs grasped with muscled arms, and decadent ghosts of Edinburgh alleys and Covent Garden stairs. He read Sherard's book on Wilde, which steadfastly declined to discuss 'the aberration which brought this fine life to shipwreck so pitiful'. Between the old world and the new, sex and death mingled, creating new myths. Most particularly, the image of the faun continually recurred in Owen's work, from 'Miners', with its 'the low sly lives | Before the fauns', to an untitled poem about the son of friends in Edinburgh: 'Sweet is your antique body, not yet young,' among 'sly fauns and trees'.

There was a knowing innocence in these shape-shifters – from Peter Pan, who would never grow up, to the teenaged C.S. Lewis's dream, in 1914, of a faun carrying parcels and an umbrella in a snowy wood. Even the boy actor Noël Coward – whom Scott Moncrieff would bring to Half Moon Street, fresh from swimming naked at Babbacombe – wrote a youthful novel about the daughter of Pan. But these hybrids, neither one thing nor the other, were also subvert emblems of a queer nature; they persisted through history like *The Tempest*, mixing myth with apprehension, if not a little fear. It is telling, given the moral outrage of the times, to note that the word panic derives from this amoral creature. Its disruptions reached back to Keats's fauns and Shelley's satyrs and Elizabeth Barrett Browning's Flush as Faunus; from Nathaniel Hawthorne's *Marble Faun* to Herman Melville's androgynous young Harry Bolton, 'a mixed being' who looked like a zebra or one of 'the centaurs of fancy; half real and human, half wild and grotesque'; and on to Aubrey Beardsley's lascivious beings, the sight of Nijinsky, a horned, piebald, cold-eyed faun falling wanking to the floor, and the future vision of a star who was half dog, a wild mutation.

Slipping between species, a boy could escape, into the sea or up to the sky. One of Wilfred's prizes from Scarborough was a bronze statuette of Hermes, god of flight, father to Pan and Hermaphroditus, a muscular form with a winged cap. The dark little deity seemed like a simulacrum of his fantastical self which he might have carried into battle if he could, with feathers sprouting out of his army cap.

Half poet, half soldier in his khaki slacks, bare torso, dark hair and slanting eyes, Owen was acting out a rite of his own duality. It was how he saw himself. In a light-hearted letter to his mother, written in the heat of the summer in France, resting in the limbo of the American field hospital, he had described his appearance: 'Clothing: sparse, almost faun.' He certainly seemed so in Osbert Sitwell's description of his small sturdy figure, with a broad forehead, wide-apart, deep-coloured eyes and 'tawny, rather sanguine skin'; young for his age, with an eager, shy air, soft, warm voice and a ready smile. He was a twentieth-century faun, picking his way across the trenches on his goat-like feet and holding not an umbrella, but a Webley revolver.

The war had not forgotten its claim on him. As the last days of 1917 turned into a new year, Owen felt confident in telling his mother, 'I am a poet's poet. I am started. The tugs have left me; I feel the great swelling of the open sea taking my galleon.' It was now certain that, despite Scott Moncrieff's efforts at the War Office to get him a home posting, he would be sent back to the front. Back in Shrewsbury, he met up with Harold. They stayed up late, talking into the night. Wilfred seemed about to take his brother into his confidence about his private life, but it was clear from Harold's attitude that any such statement would be impossible. To him, love between men was 'horrible', 'repugnant', 'revolting'. Wilfred made a joke about the supposed purity of sailors, but he had no hope of making a confession on this, their last evening together.

Pronounced fit for overseas duty, posted to a vast army camp at Ripon in Yorkshire, Owen found a river, the Ure, in which to bathe. 'I am rather weary now,' he told his sister Mary, writing at midnight in his tent, 'having Swum this afternoon, and – in consequence of the exercise –, having written a promising poem this evening.' The poem was 'Mental Cases'. He apologised to her for his exuberance. 'It comes of my recent baptism in the pleasant waters of this River. It was an amusing afternoon.'

He'd walked miles to find the bathing place, only to discover a sign reserving it 'for the Civilian Population on Wednesdays!' He decided that the civilians of Yorkshire would not be dismayed to share the river with an officer such as himself. But even here there were omens. One of the young cadets dived into the water, which was just two feet deep, 'losing his head with joy . . . and nearly lost it again. He cut it open, but as there was no brandy, he decided not to faint, and I got him safe into a cab.' Wilfred never knew that the boy later died of his injuries.

On 31 August, he got ready to leave England. Having said goodbye to Sassoon in London, and after spending the evening with Scott Moncrieff, he rose at 5 a.m., and took the seven-thirty train to Folkestone.

As I leave the station I ask a passing youth for directions to the sea. He just keeps on walking, pulling up his hoodie, declining to meet my eyes. 'I'm not from round here, mate.'

It's not difficult to see where I must ride. The streets lead down to the harbour, its arm arching out into the Channel, ending in a lighthouse. The disused boat-train station still stands, its platforms deserted, glass awning cracked, rails rusty. On either side is the sea, the only place left to go. In the distance are the chalk cliffs of Dover.

By 1918, this resort had suffered greatly. One German bomb, dropped on the street I have just cycled down, killed sixty people, mostly women and children. Folkestone was filled with refugees and soldiers. It was continually saying goodbye. Ten million troops had passed through its port; here the war had been leaking into England for four years, and England leaking into the war. You had only to step from the train and into France, twenty-three miles away. You could see it from the cliffs through a coin-in-the-slot telescope: a dark line on the horizon.

It was the last day of August. A hot, sunny day. Improbable that he was leaving England at all.

'But these are not Lines written in Dejection,' he told Sassoon. 'Serenity Shelley never dreamed of crowns me. Will it last when I shall have gone into Caverns & Abysmals such as he never reserved for his worst daemons?' He was oddly cheerful. 'I went down to Folkestone Beach and into the sea, thinking to go through those stanzas & emotions of Shelley's to the full. But I was too happy, or the Sun was too supreme.'

'I sit upon the sands alone,' Shelley had written, but Wilfred told his mother, 'my last hours in England were brightened by a bathe in the fair green Channel, in company of the best piece of Nation left in England – a Harrow boy . . .' To Sassoon, he was more revealing about this shining encounter. 'Moreover there issued from the sea distraction, in the shape, Shape I say, but lay no stress on that, of a Harrow boy, of superb intellect & refinement; intellect because he hates war more than Germans; refinement because of the way he spoke of my Going, and of the Sun, and of the Sea there; and the way he spoke of Everything. In fact, the way he spoke –'

He left his words hanging, with that boy, on that beach.

A lifetime ago another boy on another beach, Russell Tarr, had represented a new world. This unnamed Harrovian represented the last of England. He was the summation of other youths, like the 'navy boy' with whom Wilfred had shared a train compartment, golden-headed and fresh-faced, the seaman his father had always wanted him to be: 'Strong were his silken muscles hiddenly | As under currents where the waters smile.' 'And as we talked, some things he said to me | Not knowing, cleansed me of a coward-ice, | As I had braced me in the dangerous sea.' What if he had followed his father's desires, or his own? Would they, or the sea, have saved him?

He was going back to France, not for the love of his nation, but for the love of his men. 'I came out in order to help these boys,' he told his mother; to lead them as their officer, speak for their suffering. On that last day in Folkestone, Owen knew where to go: he had been here before, when he'd first shipped out from England as a soldier, back in 1916. After waiting for his shave – taking the time in the Saturday-morning queue to write a postcard home – he returned to the shore.

Wilfred, the lone wolf, on the beach.

People on the promenade, eating fish and chips, taking the air. Bands playing. Pierrots, made up in black and white, entertaining the crowds. On the streets, on trains, in canteens and shops and pubs and cafés, the civilian mixed with the military, blurring cups of tea with orders to advance. The tide coming in, and going out.

I wonder where he undressed, struggling to preserve his decency under a towel. Was he going back to war with his woollen bathers in his kitbag? The warm sun must have felt good on his tanned body, trained by war for war.

I watch him as he leaves his uniform neatly folded in a pile, striding into the sea on his short legs, feeling its rising, exhilarating chill, pushing through the waves, the water slicking back his short hair, as sleek as a selkie.

I follow him, into the surf. It's high tide, cold and green, washed with the light of the white cliffs.

For a few minutes, on that beach, under the sun, everything intensified by the light, the sea was his saviour. If he'd stayed in a little longer, everything might have changed with the next tide.

But soon he was back in uniform and boarding the three o'clock boat, bound for France.

Nearly sixty years later, eighty years old, Owen introduced me to poetry; after all, he'd been a teacher himself. In the wood-and-glass-panelled classroom of my school – run by monks in black cassocks powdered with chalk dust – our civilian English master, a harried-looking man in an academic gown, handed out the poems. Their clarion words – 'Anthem for Doomed Youth' – spoke to my teenage self-drama; they seemed to be written by someone like me, not a remote poet from the past; he was part of my resistance against the normal world. Poets could time-travel, I realised, like the starman. They also died young. In a world in which I was promised five years left to cry in, his words were ambivalent, the way I felt. I heard his voice over war cries from autumnal football pitches. I peered at his photograph, the one in all the books, its half-smile receding with each reproduction of this ordinary, handsome man.

I still cannot look enough. I had no idea of how like or unlike me he had been. I did not know him for a boy from a semi-detached house where he read his books in his boxroom, or that he walked the same shore as I did.

I turn to tell him, as he stands by my side on the beach, that he is better off, that the century isn't worth waiting for.

A few days after returning to France, Wilfred finished his poem 'Spring Offensive'. In it, he wrote of the soldiers he served; men who had come to 'the end of the world', where they 'breasted the surf of bullets'. 'Some say God caught them even before they fell'; others succumbed in the sea of mud. The dead were drowned and forgotten, as he wrote in the last line he would ever compose.

Why speak not they of comrades that went under?

The final attacks of the war were made by soldiers wearing life-jackets taken from cross-Channel ferries, advancing in the fog through flooded fields. Owen went back into battle, accompanied by 'Little Jones', his manservant pledged to protect his officer's life. Moments later Jones was shot. The two men lay together on a hillside as the servant's blood poured onto his master, 'the boy by my side, shot through the head, lay on top of me, soaking my shoulder, for half an hour', Wilfred wrote to Siegfried. 'Catalogue? Photograph? Can you photograph the crimson-hot iron as it cools from the smelting? That is what Jones's blood looked like, and felt like.'

He became his myth. Leading his men on with a new, 'seraphic' lance corporal by his side, further ahead of the line than anyone else, he roared, fighting 'like an angel', capturing a German machine gun and personally inflicting, as his citation would report, 'considerable losses' on the enemy. For this he became a hero; for this he won a medal he would never wear. He was an avenging angel, tooled up, dealing death. In the dark, chilled to the bone, another corporal produced a blanket and shared it with his officer. And before the dawn, in the hour between wolf and dog when the night sank to its coldest, acting Captain Owen led his men back under the stars, 'through an air mysterious with poison gas', recalling enough of his astronomy to assure them that it was early morning. In between the frenzied, halting action he read Swinburne, another poet sensually attached to the water. It was the only book he had left; only poetry was any good now. In another lull in the fighting, his platoon were entertained by the bizarre sight of seaside pierrots and soldiers performing in drag, all but limelit by the falling flares, like the ghosts they all already were. He had never loved his men more, and longed to tell them the war was about to end, any day now.

Four days later the company took shelter in a cottage on the edge of the Mormal Forest, which might have contained magical animals, ready for the last battle. Holed up in a vaulted brick cellar, Owen and his men planned their next attack. It was the fourth of November. Back home, boys were getting ready to light fireworks.

The Sambre–Oise canal ran through the flatlands, forty feet wide and eight feet deep, too deep to ford; a man might be lost in it. Six months before, the occupying German soldiers had jumped into this water. Naked, some still wearing their caps. Laughing. Alive. They might have been modern youths tombstoning into Torbay. This could be the Wannsee or the Serpentine. Young willows sprout from the bank.

Weeks later, as they retreated, their sappers flooded the land to forestall their pursuers. More than ever, this world was water, and its assailants sailors. The advancing British built makeshift rafts from poles and petrol cans, like lads messing about on a river. On these they tried to cross the canal, only to come under fire from young men like themselves.

Owen was last seen standing on one of the rafts before he was shot.

The dark water flowed below him, overlooked by silent trees.

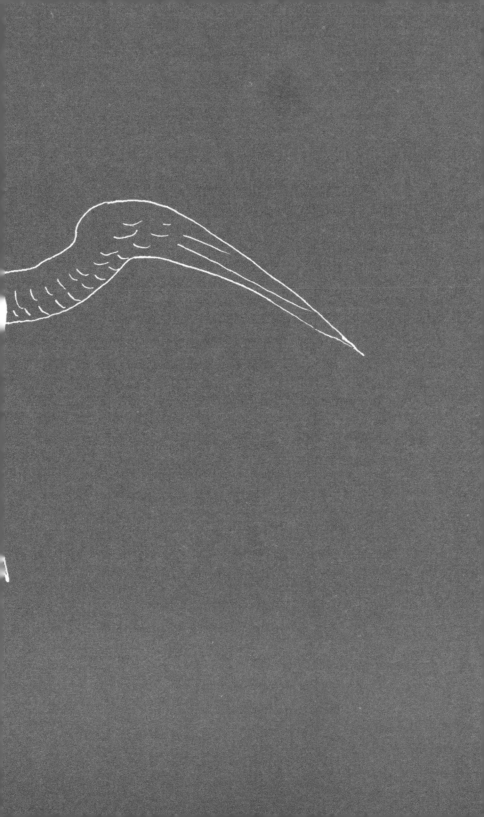

THEHANDSOMESAILOR

The branch line to Portsmouth Harbour just stops. Its rails cut off, as if the train had carried on into the sea. The station hangs over the water, its platforms resting on a rusting pier. At the far end there's a big old wooden clock which used to advise passengers of the time of the next ferry, but its hands have long since stalled.

Through the windows let into the side of the station you can look out, depending on the tide, to a seaweedy waste where mudlarks once wallowed for pennies thrown to them by onlookers; or to a swelling steely sea, buoying up the bulk of HMS *Warrior* – an ironclad, forty-gun Victorian deterrent so effective that it never saw action. Its long, low presence is now tethered impotently to the quay. Uniformed men patrol the historic dockyard beyond. This is still a working place, although its boathouses and roperies, semaphore towers and smithies, ship shops and basins long since ceased to be essential, outmoded by the ominous presence of modern warships, their sea-green superstructures surmounted by gun turrets and radar domes.

Encircled by a high wall, the dockyard is an insular citadel, with its own subsidiary isles. To the north, tucked into the harbour's inner shore, is Whale Island. It lost its nominal shape in the eighteen-sixties when it was expanded with debris dredged by convicts. What exists now, joined to the mainland by a causeway, is a functional place, surrounded by silty mud still seeded with wartime bombs; a dead zone where ships go to die and where, on family visits to Portsmouth, I'd see stranded submarines like abandoned bathtime toys. I may have been conceived by the sunny seaside, but my mother, brought on a visit to a submarine when she was heavily pregnant, nearly gave birth to me here, underwater. It's a wonder I wasn't born with a caul.

In the late nineteenth century, Whale Island became home to a Sailors' Zoo, stocked with animals that had been presented to ships' captains as gifts for the royal family. Centuries ago such beasts might have ended up in the Tower of London as part of the monarch's menagerie, but now their fate had fallen to men whose wandering had turned their hearts sentimental. According to a 1935 edition of *The Times*, the zoo 'grew out of the sailor's fondness for pets of all sorts, and the care he gives to their well-being'. Somewhat worryingly, it also overlooked the parade ground of the island's gunnery school, as if its inmates might provide exotic target practice.

It was a reflection of the navy's secret love of eccentricity, bred by its romance with the sea. 'For lions and other animals there are spacious iron cages, much like those of the Zoological Gardens. The marsupials have large grass paddocks to roam in; for the aquatic birds there are big ponds, and there are large aviaries for birds of other species.' It must have been odd for locals to hear the island's strange noises from over the water, as if a little bit of Africa or the Arctic had been towed into the harbour. The collection was a growling, squawking index of colonialism, complete with a Shakespearean bear pit. 'Among its first carnivora', notes *The Times*, was a polar bear named Amelia, given by Inuit people to HMS *Grafton* in 1904. Her fellow inmates included one of Captain Scott's sled dogs who had done brave duty in the Antarctic, and a parrot named Calliope Jack, the sole survivor of HMS *Calliope* after it sank in a Samoan storm. Jack was the zoo's veteran: he lived for thirty-nine years on Whale Island, and died in 1919. Another resident, Tirpitz the pig, was rescued from a German warship off the Falklands in 1914. Kept onboard as food, he had been left below when *Tirpitz* was scuttled, and managed to make his way to the upper deck. He swam for an hour towards a British ship, one of whose officers nearly drowned trying to save the frightened animal. Tirpitz too spent the rest of his life on Whale Island, before performing his final service on a dinner table.

The island had become an animal League of Nations, a melancholy ark. In 1940, its last captives met an abrupt end: on 27 May, lionesses Lorna and Topsy, polar bears Nicholas and Barbara, and sun bears Henry and Alice, were summarily shot. It was a logical move. Apart from fears that the animals might escape and terrorise the city, the country as a whole anticipated food shortages and air raids during which, civilians were informed, no pets would be allowed into the shelters, nor would they be included in rationing. Along with the non-human occupants of Whale Island, domestic animals became the first casualties of war. All around Britain, owners gave up their dogs and cats to be put down. In 1939 – within a week of the publication of an official government pamphlet advising 'it really is kindest to have them destroyed' and recommending the '"Cash" Captive Bolt Pistol' as 'the speediest, most efficient and reliable means' of doing so – three-quarters of a million pets were shot or gassed, long before the first bombs fell.

There are no caged animals in Portsmouth Harbour now. Rather, centre stage in the port's stalled drama stands the single most significant historical artefact in Britain: HMS *Victory*, which I boarded that morning, for the first time since I was a boy.

Victory still stands close to the sea, but unlike *Warrior* it lacks the reassurance of the waves lapping its hull; its magnificence is marooned in a dry dock. This morning part of its prow is rudely exposed, the bride stripped bare of layers of paint, subject to constant conservation work. The ship's regal state is somewhat diminished by this care and attention, like a proud octogenarian

303

forced into a nursing home; and by the fact that it has been up-staged by a third warship, one which didn't manage to stay afloat at all: Henry VIII's flagship, *Mary Rose*. After marinating in the Solent for four centuries, the wreck was hauled from the turbid water, and now lies under cover in climate-controlled gloom, its splintered timbers slowly being sucked dry of salt water as their expanded cells are pumped full of silicone. If this is surgery on a grand scale, then the low hum of machinery in the darkness lends the extended shed the air of a giant intensive-care unit.

Arrayed around her – I fall into the gendered terms, seduced by these ships, as much mothers as mistresses, and their crews as lost boys – are the objects the drowned crew left behind. They're displayed in faintly-lit glass cases, like one long aquarium. I'm peering into the aftermath of disaster: from the surgeon's velvet skullcap to personal sundials, Tudor wristwatches; from immac-ulately preserved longbows made from French yew to skeletons of the six-foot archers who used them, their overdeveloped right arms testament to their profession. Peter, my archaeologist friend, points out grooves where strong muscles were attached, and jaws where the bone grew over empty tooth sockets so smoothly that in some the entire lower mandible became one gummy structure.

These objects are dumb, but they speak of the last moments of five hundred men, their names lost to the sea. A hide jerkin which once spread over the substantial chest of its anonymous owner was impressed with his ribs as he came to rest in the mud, leaving his shape in leather. In a nearby case is another nameless skeleton: the ship's dog. The label supposes it was kept to control the rats foraging on scraps dropped by sailors, but it is easy to imagine how it was loved, too, for its own sake.

Given the contemporary attention lavished on this Tudor wreck – the sort of ship that might have foundered off Prospero's isle – her eighteenth-century counterpart seems vaguely forgotten, as if she'd been wheeled out of the ward and left to fend for her-self. Dismasted and deprived of her figurehead, *Victory* is naked, exposed to tourists' stares and all-seeing smartphones. And where *Mary Rose* is plumped up with silicone, *Victory* is continually rebuilt in her own image, much as we are always regenerating ourselves. Her wounds heal over; her ship's knees are as scarred as my own.

As I climb the gangplank – a route once restricted to officers with gold braid around their sleeves – I feel underdressed in my shorts and sandals, expecting someone to bark at me, ordering me to adjust my attire.

Ducking under the doorway, I bend to enter the hallowed maw of what is as much sacred architecture as historic ship. This revered interior, a lowered cathedral, requires genuflection. The layered decks demand obeisance to their holy timbers; the headbutting beams force me to bow. Most historic sites forbid you to touch, but here you cannot help it. You are in constant contact with the structure with hands and head and feet, perpetually aware that you are being admitted to a shrine.

'It's two hundred and fifty years old,' an American woman tells her young son as they ascend ahead of me. It's not clear whether this is a remark or a reprimand.

Winding through the wooden aisles towards the stern, we're drawn into a burst of light like a nave. In Nelson's Great Cabin, a personable, eager young rating in white shirt and black slacks (a term for which such an item is made) stands at ease, legs apart, behind the admiral's table – a position which, two centuries ago, he could have occupied only if he were waiting on his superiors or awaiting sentence. Today's handsome sailor imparts his information on a rolling, need-to-know basis, repeated with renewed enthusiasm for every new arrival.

The details are anecdotal, engaging. He tells us all about the en suite facilities the admiral had at his disposal; how he preferred to sleep in a specially designed low leather chair because his constant struggle with seasickness made his swinging cot an uncomfortable resting place; and how the entire interior was transformed for battle, its panels and partitions folded away and cannon rolled in as it became a war room.

All this is delivered with cheery briskness. I feel I want to take a breath for our guide, but he clearly enjoys his performance; he is, after all, flanked by superior set-dressing. In one corner stands a tall vitrine, just shy of the height of our instructor. It holds a headless dummy draped in a replica of Nelson's full-dress uniform: a blazing, glamorous get-up, from his cutaway coat and red silk sash to the diamond *chelengk* on his cocked hat, plucked by Selim III, Sultan of the Ottoman Empire, from his own royal turban and awarded to the admiral for having won the Battle of the Nile. Stuck in his bicorne like a comet caught in its felted brim, its fêted recipient was the only infidel on whom this starry decoration had been bestowed. The outrageous assembly of eighteen diamonds – then valued at eighteen thousand pounds – was animated by a clockwork mechanism which slowly revolved its centrepiece, all the better to catch the light, while its thirteen wired rays – each representing an enemy ship defeated in the battle – trembled like studded feathers. Nelson loved to wear this 'plume of triumph', at a time when it was not usual for an officer to bedeck himself so. Sir John Moore, the British general, met Nelson in Naples shortly after the battle and gruffly described him as 'covered with stars, medals and ribbons, more like a Prince of Opera than the Conqueror of the Nile'.

The glittering commander-in-absence illuminates the space, casting glorious rays on the proceedings. Far from the gloom of the rest of the ship, his day cabin is flooded with light: the stern is one long wall of glass, a giant Georgian bow window. The walls are delicate duck-egg blue; the floor canvas, painted to resemble the black-and-white tiles of an elegant interior. It might as well be a Regency drawing room as a working war office; a suitable setting for a Prince of Opera covered with stars.

But out of its princely glow, the mazy darkness returns, brown and murky. I climb on through the ship, feeling my way along the decks, as disorientated as if I were lost in a multi-storey car park; only here the floors are laden with three-ton guns rather than Fords or Audis. *Victory*'s sides are spiky with cannon which once disgorged a dragon's breath, each discharging a pulverising death. The entire ship is an organic war machine, almost animal itself, looped and bound with hemp and canvas and wood and iron, soaked to the skin with tar and blood and sweat and piss. If a whale ship was slick with spermaceti oil, then this vessel, this container, was lubricated with the oil of humans.

I'm funnelled through midshipmen's cabins with their neat cupboards and plank bunks, past the galley – the only place on the ship where fire was allowed, in an oven the size of a carriage – down to the whale-belly bilges laden with pig iron and what looks like railway ballast, laid to steady this lurching leviathan. I'm led by an invisible, smelly sailor, head and body bent, along gangways and alleys and into dead ends where all manner of misdemeanours and conspiracies might have taken place while the officers were busy in their drawing room above, planning their next engagements over a glass of port. Finally, bending ever lower, I reach the place where he lay. It is marked by a gilt-framed copy of Arthur William Devis's painting on an easel, displaying, a still from a Regency movie, a re-enactment of the admiral's death.

Devis, an enormously successful but reckless artist who was let out of debtors' prison to create his canvas, was allowed to board Victory on her return from Trafalgar so that he could sketch the principal players, both living and dead. Like Trelawny, he'd had his fair share of drama. He had served on an East India ship and survived an attack by New Guinea islanders during which he was wounded in the face, leaving him with a locked jaw. During his career he had reached great heights – he was paid an astonishing £2,530 for his portrait of the governor-general of India, General Charles Cornwallis – and great depths, having been imprisoned as a bankrupt. Now came Devis's last chance to make a mark on posterity – and pay off his debts. His commission was the most prized in England. He would have to do it justice.

The artist stayed on the ship for a week, sketching every aspect of the scene with forensic detail, and when Victory sailed to Chatham for repairs to its own battered carcass, Devis accompanied it. He assisted William Beatty – the ship's surgeon who had attended the dying Nelson – as they decanted the admiral from the alcohol-filled cask in which he had been pickled. Manhandling him like a giant fish, the two men dressed the hero in his uniform, preparing him for his final performance.

The means of preservation had left Nelson's features distorted and discoloured, and despite Beatty's attempts to restore them by rubbing, it was decided that Nelson looked too unlike Nelson, and unfit for public display at his forthcoming lying in state. Yet for Devis, it was an extraordinary opportunity. He could do what a doctor could not, restoring the dead with a gaudy slap of paint.

The finished canvas was, as Sir John Moore might agree, a suitably operatic scene. Nelson has been felled from above, and lies ashen and limp, his life ebbing away. His accoutrements and awards are discarded uselessly beside his body, which even now looks greenish, as if already steeped in fine brandy and sweet wine; his clothes lie at his feet, the surgeon having cut away his breeches to allow access to his wounds. Nelson was perfectly aware of the significance of his death – from the handkerchief placed over his

face and medals as he fell so that his men should not recognise him, to his comment to Hardy, 'They have done for me at last. My backbone is shot through' – to his urgent request that his body should not be thrown overboard, as other bodies were being tossed into the Atlantic all around him.

Around his pale glowing flesh – 'a dismal light about it, like a bad lobster in a dark cellar' – gather mechanical figures. Like Rembrandt's depiction of a dissection or a resurrection by Titian, Devis's painting plays tricks with dark and light; it manages to be both overlit and obscure at the same time.

His rival, Benjamin West – a Pennsylvanian Quaker, now president of the Royal Academy, and celebrated for his spectacular *Death of Wolfe* – complained that Devis's work was not an 'Epic representation', and that Nelson looked 'like a sick man in a Prison Hole'. (Devis might have countered that at least he knew what such a hole looked like.) Given that West had painted his own version of the most famous event in British history, the contrast between the competing works – these aesthetic autopsies, both fighting for historical supremacy – is a wonder to behold.

West presents an action-movie version of Nelson's demise, all red-coated marines and a pair of bare-chested sailors who wouldn't look out of place fighting in a dance hall in Pompey tonight. Hats are raised, heroic stances struck. Every earnest, seaworthy face is turned towards the stricken admiral, who sinks gracefully like a stove schooner in his officers' collective embrace. Meanwhile the battle rages behind, triumphantly won even in its victor's fading light. To the boy in me, such scenes were thrilling. In my early teens I was obsessed with the Napoleonic Wars; I spent hours painting tiny metal soldiers with the peacock colours of hussars, chasseurs, dragoons, grenadiers, cuirassiers, lancers and mamelouks, their frogging, plumes, epaulettes, dolmans, pelisses, shakos and leopardskins the epitome of the exotic military dandy.

How much darker is Devis's vision! Where there's scarlet and gold in West's painting, there's murk and gloom in Devis's. His is the mumblecore version of Nelson's death, all whispers and shadows, imbued with mortality rather than pomp. One is a celebration, the other a Christian parable, a rousing crescendo or a plangent aria: both issue from a theatre of war which neither artist witnessed, staged on an immortal ship with its sails as backcloths, its crew as the angelic chorus, and its rigging as the stairway to heaven.

It's all getting a bit fetid down there, fuggy with all this smoke and blood and guts, and I'm relieved to return to open air, ascending to the most sacred spot of all, the ground zero of British history. Screwed into the top deck is the plaque marking the place where the sniper's musketball did its job, entering Nelson's left shoulder to lodge between the sixth and seventh vertebrae, carrying golden threads from his epaulette as it travelled on its fatal trajectory, and skewering the heroic corpus in an Olympian apotheosis, forever falling back, over and over again.

HERE NELSON FELL
21st Oct 1805

Except that this is not where he fell, nor is this the 'original' plaque which once marked the spot. Like much of the ship, the quarterdeck has been replaced even since I last stood here. And this morning the timbers are being laid again.

'What are they made of?' my companion – who happens to be named Horatio – enquires of one of the workers, who is busy kicking a black bin bag full of rubbish over the place where he lay.

'Wood?' the man replies, pleased with his own sarcasm.

Not even hearts of oak; as Horatio points out, the new decking is comprised of tropical hardwood.

Suddenly, I feel fooled. Is any of this real? The cannon ranged along the lower decks have long since been replaced by fibreglass replicas, for reasons of weight and wear, we're told; although the cannonballs are made of concrete rather than iron. The entirety of this '*Victory*' might as well be a set for a seventies television production, with a different crew: a director in thick-rimmed spectacles and sideboards, lumbering cameras on wheels wielded by men in huge headphones, and a smart young continuity woman, clipboard at the ready. Someone plays some stirring music.

When the Athenians honoured Theseus' ship by replacing its rotting planks, their philosophers agonised as to whether it was Theseus' ship any longer, any more than a river is the same river as its waters constantly flow out to sea, or than you are the same you since your body has rebuilt itself many times since you came into being. We are memory, not history. At what point, in the transmigration of maritime souls, did *Victory* stop being *Victory*, if it ever did? For the long nineteenth century the Regency relic lay along the quayside, accessed by Victorian visitors by boat and ladder, reincarnated in the Victorian image like an over-restored country church; a Victorian version of what it should look like, rather than the Georgian ship Nelson had known. In 1922 it was brought into dry dock as a pageant, a nineteen-twenties edition of itself. In the Second World War it was a symbol of resistance while the dockyard buildings burned around it in the Blitz. Then it became a life-size Airfix kit for boys like me.

Now, in the twenty-first century, it is no more or less *Victory* than a forest is the sum of its replanted parts. In a Hampshire nature reserve, another plaque tells me that the oaks from which *Victory* was made took three hundred years to grow, three hundred to flourish, and three hundred to die. Perhaps the ship should be allowed the same dignity. In the interests of historical accuracy she ought to be left to rot there in the dockyard as a memorial to all that she meant, and now does not mean.

On Christmas Eve 1849, wearing his new green box coat, of which he was inordinately proud, Herman Melville took a cab from his lodgings in Craven Street, which lay just off Trafalgar Square and its lofty statue of Britain's naval hero. The narrow road ran down from The Strand towards the unembanked shore, where the Thames flowed wide and filthy, 'the inscrutable riverward street packed to blackness', as Henry James would see it. After three months away from home, Melville was leaving London.

After a five-hour journey from Waterloo station, he arrived at Portsmouth, where he spent the night at the Quebec Hotel on the Point, a spit of land jutting out into the harbour. Nicknamed Spice Island, it had long been a lawless place, a huddle of taverns, warehouses and whorehouses. The hotel still stands, a genteel Georgian building, and its address – Bath Square, at the end of Bathing Lane – advertises its original function: as a bath house fed with salt water, the waves almost running into its dining room and up to its guests' tables. Predating Torquay's medical baths by half a century, the Quebec was a testament to a serious pleasure: that of entering the sea for its own sake.

The building is dwarfed on either side by stone ramparts. At their feet runs a slender beach where, in summer, old men with blurred tattoos turn the colour of coffee, and where local lads throw themselves off walls made to repel invaders, their pale bodies arcing into the water like seals. I swim there too, in the narrow channel dredged deep enough to accommodate aircraft carriers.

I imagine Herman watching me from his room, before taking to the streets.

It was late December, and a cold wind blew up the Solent. That coat came into its own. This was the sort of place to which Herman was always drawn, like the Battery in Manhattan, where he was born; places where the streets stopped and the sea took over. On his early-Christmas-morning stroll, he 'passed the famous "North Corner"', as his journal records. 'Saw the "Victory", Nelson's ship at anchor.' Then he returned for breakfast, only to be rudely interrupted by news that the ship which was to take him home had appeared in the harbour. In a flurry of capsized coffee cups, he dashed to grab his bag, and set off on his voyage back to New York.

There is no hint, in his scant account, of Melville having actually boarded *Victory*; it was then a receiving and training station, not officially open to visitors, although 'this would not prevent an enterprising sailor extending an informal invitation to come aboard for a few shillings'. But he certainly boarded it in his imagination: when Father Mapple delivers his sermon on Jonah from a pulpit-prow in *Moby-Dick*, 'impregnable in his little Quebec', a ray of light, shed from a window with an angel's face in it, illuminates a spot on the chapel floor, 'like that silver plate now inserted into Victory's plank where Nelson fell'. Nor is it any wonder that this story so affected the young American on his visit to Portsmouth, since the shore from which he left England was the same from which Nelson embarked for his fated appointment.

On that last day on land, the vice admiral had tried to 'elude the populace by taking a byeway to the beach', only for a crowd to gather at Southsea, 'pressing forward to obtain a sight of his face; many were in tears, and many knelt down before him, and blessed him as they passed'. It was as if he were already a saint.

The numbers swelled as they reached the sea wall, pushing towards the parapet, 'for the people could not be debarred from gazing till the last moment upon the hero – the darling hero – of

England'. At two o'clock in the afternoon, Nelson embarked from the bathing machines, and was rowed to Victory past the wheeled huts lined up on the shore.

A month later, on 21 October 1805, he strode on deck to face the enemy. In the literary accounts that match Devis and West's paintings with their florid words, he was a standing target. According to the poet Robert Southey – on whose biography of the hero Melville drew – Nelson 'wore that day, as usual, his admiral's frock-coat, bearing on the left breast four stars of the different orders with which he was invested. Ornaments which rendered him so conspicuous a mark for the enemy were beheld with ominous apprehension by his officers.' For his part, Melville saw 'a sort of priestly motive' which led Nelson to 'dress his person in the jewelled vouchers of his own shining deeds; if thus to have adorned himself for the altar and the sacrifice'. In fact, that day Nelson wore his undress coat with sequin replicas of his awards sewn over his heart. Yet those paillettes and silver-gilt thread did indeed turn his coat into a conspicuous costume, 'an ornate publication of his person'.

These stories preoccupied Melville, who was busy creating his own myths. In London, he had made a pilgrimage to Greenwich, taking the steamer from the Adelphi down the Thames; he was already familiar with the river, having crossed from Wapping by the new tunnel to Rotherhithe, 'flinging a fourpenny piece to "Poor Jack" in the mud'. He even claimed to have seen a beggar on Tower Hill with a board hanging around his neck like an albatross – only it was painted with the whale that had bitten off his leg.

Greenwich's naval hospital was well known for its veterans, identified by their own wooden stumps and archaic blue frockcoats and cocked hats which earned them their nickname, Greenwich geese. The building itself was a great ship beached on the riverbank, with figureheads on its walls and peg-legged sailors accommodated in 'cabins' along with their mementoes, from maritime paintings to stuffed seabirds.

Wandering along the terrace, Melville met a fellow American, 'an old pensioner in a cocked hat with whom I had a most

interesting talk ... a Baltimore Negro, a Trafalgar man'. This
chance encounter would assume a certain importance for the
writer. During the Napoleonic Wars, many Americans were press-
ganged into the British navy, 'all kinds of tradesmen and Negroes'
– twenty-two on *Victory* alone. The Baltimorean told Melville that
even the gaols were raided for crews.

In a photograph taken in 1854, five years after Melville's visit,
a group of Greenwich veterans sit on a bench. Among them is an
elderly black man. Records indicate that one Richard Baker, born
in Baltimore, served at Trafalgar on HMS *Leviathan* and later joined
the pensioners; his ship was retired to Portsmouth Harbour as a
prison hulk, holding convicts bound for Van Diemen's Land. Was
this Melville's countryman? He sits in antediluvian company, an
exotic figure clutching a cane, a medal on his waistcoat, his hair
turned entirely white.

To Melville, this sailor, an alien like himself, was a living
memory of a legendary past, in a place where time and space began

and ended: embedded in the hill behind was a brass line over which one could step from one hemisphere to another. When he walked into the Painted Hall, Melville found fifteen hundred veterans at dinner; the contrast between the rough mariners in their frockcoats and the baroque interior was remarkable: 'Pensioners in palaces!' And in an anteroom he came into the presence of the hero himself: a row of glass cases filled with Nelsoniana.

Six years later, in 1856, Melville's friend Nathaniel Hawthorne would also visit Greenwich. He'd just arrived in London from Southampton, where he'd seen Netley Abbey and was fascinated by the gypsy woman who'd lived in the ruins for thirty years. Ever attuned to the gothic, Hawthorne was struck by the Painted Hall and its sanctuary, 'completely and exclusively adorned with pictures of the great Admiral's exploits. We see the frail, ardent man in all the most noted events of his career, from his encounter with a Polar bear to his death at Trafalgar, quivering here and there about the room like a blue, lambent flame.' For Hawthorne, as for Melville, Nelson was a legend. 'But the most sacred objects of all are two of Nelson's coats, under separate glass cases.'

One is that which he wore at the Battle of the Nile, and is now sadly injured by moths, which will quite destroy it in a few years, unless its guardians preserve it as we do Washington's suit, by occasionally baking it in an oven. The other is the coat in which he received his death wound at Trafalgar. On its breast are sewed three or four stars and orders of knighthood, now much dimmed by time and damp, but which glittered brightly enough on the battle-day to draw the fatal aim of a French marksman. The bullet-hole is visible on the shoulder, as well as part of the golden tassels of an epaulet, the rest of which was shot away. Over the coat is laid a white waistcoat with a great bloodstain on it, out of which all the redness has utterly faded, leaving out of it a dingy yellow hue, in the threescore years since that blood gushed out. Yet it was once the reddest blood in England, – Nelson's blood!

Seventy years and many wars later, these relics retained their power. In 1926, when writing *To the Lighthouse*, Virginia Woolf visited that Greenwich chamber and found it heady with emotion. The sight of the coat whose decorations Nelson had hidden with his hand as he was carried down, 'lest the sailors might see it was him'; his 'little fuzzy pigtail' tied in black; and his long white stockings – 'one much stained'; all these prompted her almost to burst into tears, '& could swear I was there on the Victory'.

In the Ordnance Survey of Great Britain, the Thames shore is accounted part of the country's coast, the sea inside the city turned inside-out. Up until the nineteen-thirties the riverbank at Greenwich was used as a beach from which Londoners swam, as if it were a resort. Walking there a few years after Woolf's visit, Denton Welch passed the desultory strand where the low tide still reveals blackened bones, empty oyster shells and the soft stems of clay pipes like the debris of some long-forgotten feast. On the shore he saw a pair of children playing 'some spiritualistic ghost game ... close to the horrible black water'. The young artist felt overwhelmed by the insistent sights and sounds of the river, 'their never-ending story of time passing, longing, death'. Later, riding over the Thames on a bus at night, he looked down from the bridge into the swirling darkness and imagined how awful it would be to swim around its stone piers. He wondered if they had barnacles on them, and thought of the black mud at their edges, so deep that he might sink up to his neck in it. What if he were to see someone jump from the parapet? Would he dive in too, only for the suicide to clutch at him frantically, drowning him as well?

I realise, as I walk along the river, that my pockets are so full of stones I've collected from other beaches that if I fell in I'd probably sink as quickly as Virginia sank into the Ouse. One step over the embankment and there'd be nothing the modern world could do to save me; I would enter another world. The solidness of the stone edge only intensifies the intimation of mortality: when it replaced the beaches of the city, the embankment made the Thames flow more dangerously, 'running away with suicides and accidentally drowned bodies faster than a midnight funeral', as Dickens wrote. Its blocks still stand as gravestones for the lost, the two dozen or more souls who throw themselves into the river each year as it courses through the capital.

Behind me rise Wren's colonnades and cupolas – 'among the most splendid things of their kind in Europe' – measuring out the imperial reach. This is London's water palace, built to service its maritime empire. And at the heart of the baroque complex, so little changed that it is often used in films as a stand-in for the eighteenth-century city – is its chapel, its domed entrance draped with semi-naked female figures and captioned with Biblical exhortations for those who would take to the sea.

WHICH HOPE WE HAVE AS   FAITH IS THE SUBSTANCE
AN ANCHOR OF THE SOUL   OF THINGS HOPED FOR
BOTH SURE AND STEADFAST   THE EVIDENCE OF THINGS
                                                        NOT SEEN

To one side stands a memorial dedicated to Sir John Franklin and his expedition, listing the crew lost on his ships *Terror* and *Erebus*. Overlooked by shards of marble ice – from which, as Woolf imagined, the explorers looked out 'to see the waste of the years and the perishing of the stars' – a disconsolate figure in mittens and boots mourns one of the lost, found by an American expedition in 1869 and whose anonymous remains are interred below, an unknown warrior of the Arctic.

Leaving this sombre interior, I cross to the Painted Hall and walk into a bursting polychromatic lightbox. Floodlit by huge windows, its vast murals are as vivid as a colonoscopy. This echoing space, where Melville watched the inmates dine, was a truly extravagant interior for such an ordinary purpose, created by the architects Wren and Hawksmoor and the artist James Thornhill. Its opulent canopy, a Technicolor trick of the eye, is animated with classical gods, monarchs and emblems of the four known continents – Australia was yet to be discovered – with New England represented by a Native American in a war bonnet; soaring across the walls are cosmological symbols of the seasons and stars. Washed by grey London light, the effect of this panoply is overpowering, sending the visitor below scurrying across the black and white tiles like a beetle.

And at the point at which any sacred place would be dedicated to a saint there are two brass plaques let into the floor. One commemorates Vice-Admiral Collingwood; the other, his fellow commander. For three days in January 1806, thirty thousand souls trooped into this chamber, its windows boarded up and its murals deadened with black crêpe, lit by hundreds of candles. It was a sepulchral stage set, directing all eyes to the hero whose disembowelled corpse lay in a black-velvet-covered coffin studded with bronze symbols of a sphinx, a crocodile and a dolphin; heraldic familiars from his victories to accompany him into the afterlife.

The sheer press of people threatened to throw the arrangements into disarray; the authorities feared a riot. On 8 January

Nelson's coffin was placed on a barge and rowed up the Thames to the Admiralty, accompanied by so many vessels that one could have stepped across them from one bank to the other, and passed under bridges so crowded that bystanders fell off and drowned as a result of their eagerness to see the last of their saviour.

Brought to shore at the Whitehall Stairs, the bier was carried on a funerary car shaped to resemble *Victory*, complete with glazed stern and a figurehead bearing a victorious wreath. The procession was so long that its end hadn't left Whitehall by the time its beginning reached St Paul's; all the while, it was watched by murmuring crowds. Inside the cathedral, another Wren interior, a huge tiered wooden arena had been built to accommodate the congregation. The sound of military bands playing a tune to Psalm 104 – 'Yonder is the sea, great and wide | which teems with things innumerable, | living things both small and great. | There go the ships, and Leviathan which thou didst form to sport on it' – swirled around the Whispering Gallery, which was lit by one hundred and sixty Argand lamps burning whale oil.

Then, in a final piece of theatre which to one newspaper smacked of a 'stage trick', the coffin was lowered directly through the floor of the nave to the crypt below. Sir John Moore would have harrumphed. At that point, the sailors of *Victory* came forward. They were supposed to fold the colours and lay them on their commander's coffin, yet such was their fervour that they tore the flag into pieces, the scraps of red, white and blue becoming relics of a man who had been translated into a myth.

Meanwhile, a few miles away, William Blake was at work on his own mystical tribute – *The spiritual form of Nelson guiding Leviathan, in whose wreathings are infolded the Nations of the Earth* – in which the admiral rises naked and transcendent from the dead, an Apollo arrayed on the coils of a sea monster.

Three years later, as Blake displayed his finished picture in his brother's Soho shop, the river received another body. In 1809 a 'wonderful large fish' was seen south of Greenwich at Sea Reach.

This leviathan – a seventy-six-foot-long fin whale – was shot, taking four hours to die, before being displayed at Gravesend: a suitably-named site, where the grey mud is the colour of a Weimaraner, where Pocahontas died of disease before she could return to Virginia, where *Mayflower* dropped down with the tide on her way to Southampton, and where Franklin's Arctic explorers would attend their last service at a waterside chapel. Thousands viewed its carcase.

It seemed each generation summoned its own sacrificial beast up the river. In October 1849, Melville's arrival in London was preceded by another whale, laid out on the front page of the *Illustrated London News* below a report on the Irish famine, as if he had imagined it out of his future story. Labourers had seen something dark floating on the water, 'when suddenly the violent plunging and dashing of one end of it intimated to the men that it was some living monster of the deep'. The fifty-eight-foot fin whale 'made desperate efforts to obtain its freedom', but was duly lashed with ropes and dragged onto the beach where, 'by the aid of a sword its life was dispatched, and the men then set about inclosing it for exhibition'.

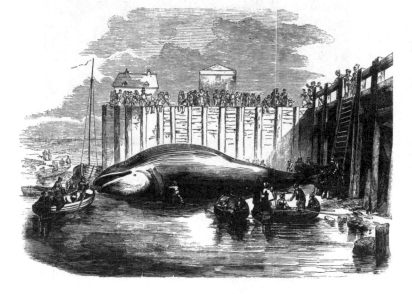

The rain is pouring down at the end of an Indian summer as I cycle into the car park off Trafalgar Road, Greenwich. An anonymous warehouse stands before me; it might contain frozen foods or car parts. I sign in, deposit my soaking rucksack, and am conducted by Amy and Louise into a cavernous, climate-controlled space. Amy, who is from Pennsylvania, is lamenting London's lack of an autumn. She strides to one of the giant racks and pulls out a wire grid on wheels. Hanging from it rather haphazardly, as if it had just been found in a skip, is a large gilt frame; and within the frame, a full-length portrait of Nelson painted by Leonardo Guzzardi to commemorate the Battle of the Nile in 1798. It was this victory which earned Nelson his *chelengk* – along with enough booty to please any pirate, including another diamond-set tribute from the Russian czar, a sable pelisse from the grateful monarch of Sardinia, and the dukedom of Bronté from the King of Naples.

In Guzzardi's painting, commissioned for the sultan in return for his gift, Nelson stands resplendent in dress coat and white breeches, as bedecked as a Christmas tree. One arm points out of the frame to the enemy fleet he has destroyed. The other, of course, is lacking; his empty sleeve lies across his breast, both useless and potent. His stance is balanced; as the two curators note, it is that of a dancer rather than a fighter. 'Everything in the eighteenth century was about the legs,' Amy tells me. To modern eyes, the hero's shoulders slope inordinately; all attention is focused downwards, in an unmistakeably sensual manner.

I see an entire era anew. It's like being given access to a secret code, only to realise how obvious it was all along.

Nelson slips one slender leg in front of the other. His pose reminds me of Mr Turveydrop, the dancing master in *Bleak House*, a pomaded relic of the Regency, usually to be found 'in a state of deportment not to be expressed'. The classical ideal, says Amy, was to look like a marble statue.

But all this civilian distinction, more suited to a ball than a battle, is dispelled by Nelson's face. Only in this portrait are his wounds so visible; it is more pathology than painting, like a medical

illustration from the Royal College of Surgeons. Running from his hairline to his eyebrow (half of which has been shaved off) is an insane Ahabian scar – as if he'd been struck by lightning – a zigzag rip which left a flap of skin dangling in front of his already blind eye, exposing his skull.

This living Nelson is a damaged, emaciated, asymmetrical figure. As the skin fell over his face during the battle, blood pouring into his eyes, he collapsed into the arms of Captain Edward Berry and was carried below, crying, 'I am killed, remember me to my wife,' and calling for his chaplain. Stitched by the ship's surgeon in dim light, the repair is not an invisible one. Any dandy would have reprimanded his tailor for such work.

Nor was it the first time he had nearly died. In an expedition to the Arctic as a fifteen-year-old coxswain, Nelson had leapt onto the ice to shoot a polar bear whose pelt he wanted as a present for his father. In the legendary scene, he discharged his weapon, missed, and only a chasm opening up between him and the beast preserved him. 'Never mind,' Southey had him say; 'do but let me get a blow at this devil with the butt-end of my musket, and we shall have him.' Nelson was, said Southey, feeble in body but affectionate in heart. In India he fell victim to a disease which defied diagnosis and left him 'reduced almost to a skeleton', unable to use his limbs. In the Caribbean, having avoided being bitten by a venomous snake, he drank poisoned water that left 'a lasting injury upon his constitution' according to his doctor; perhaps he was a victim of the same vodou which poisoned Bro's moonlit dolphin. He then contracted dysentery, which rendered him so helpless that, like Elizabeth, he had to be carried to and from his bed, 'and the act of moving him produced the most violent pain'. Even his family seemed prone to dramatic tragedy: in 1783 his sister Anne 'died in consequence of going out of the ballroom at Bath when heated with dancing'.

Nelson was so attuned to his own mortality that he had a coffin set upright in his cabin behind his dining chair. He was forever rehearsing his noble fate, falling and rising and falling again. He was one long wreck. The year before the Battle of the Nile, when his arm was smashed by a musketball at Tenerife, he told his sixteen-year-old stepson Josiah standing next to him, 'I am a dead man.' After it was amputated, his arm reappeared as a phantom limb: feeling his own ghostly fingers pressing into his palm, Nelson declared he'd discovered 'direct evidence of the existence of the Soul ... For if an arm can survive annihilation why not the whole person?' In *Moby-Dick*, as he lies sleeping with Queequeg's arm around him, Ishmael is scared by the childhood memory of his sleep-deadened hand; while Ahab, feeling the tingle of his own lost leg, muses on the absence: 'How dost thou know that some entire, living, thinking thing may not be invisibly and uninterpenetratingly standing precisely where thou standest; aye, and standing there in thy spite? ... And if I still feel the smart of my crushed leg, though it be now so long dissolved; then, why mayst not thou, carpenter, feel the fiery pains of hell for ever, and without a body? Hah!' The age invented Nelson: a Creature of his constituent parts; or already a resurrected god, as Blake saw him.

Amy and Louise are busy on the other side of the room. Working a sort of windlass, Amy opens up a floor-to-ceiling unit and, with Louise's help, takes out three white body-length bags, one after the other. Carrying them as carefully as orderlies might lift a patient or a cetologist a dead porpoise, the two women lay them on the table. One by one, their shrouds are unwrapped for my benefit.

Out of that bright whiteness, in the fluorescent-lit laboratory interior, the god is revealed in wool and thread and metal. The intense, almost black navy cloth is revealed under the kind of light it would never have seen in its occupant's lifetime, its lustrous darkness enhanced by rows of gold buttons two centuries have failed to dull. The mattness of the material, which lies so flat, is heightened by these gilt bursts. It is in these contrasting qualities that the essential power of these garments lies. Over the next two hours we talk over the three coats like doctors at a bedside, moving from one to the other as our assessments are made. Every detail, every loop of gold thread, every delicate stitch is discussed and diagnosed as we admire artefacts as finely engineered as any ship of the line.

Opening up the first – the undress coat worn by Nelson at the Battle of the Nile, as seen by Melville and Hawthorne two centuries ago – Amy shows me the fine linen lining, darkened at the neck and lower down by rusty brown stains. Nelson's blood, she says matter-of-factly, even though she is as aware as Hawthorne, her fellow American, was of the ritual power of those words. Flipping the coat over with Louise's help, like a nurse administering a bedbath, Amy points out a greasy horizontal smear on the back, just below the collar. It is the tideline of Nelson's pomade, left behind as his pigtail brushed back and forth while its owner issued commands and exhortations. This honoured raiment is a grubby, scabby rag, a reminder of the messy business of being human. It would normally be Amy's signal duty to expel any organic substance from such an important item. Here it is her task to preserve Nelson's memory and conserve his genetic residue, as if he might at some future point be regenerated when England needed him again.

But then, the cloth itself is impregnated with the subtle oil of Spanish Royal Escurial merino sheep, their fleeces coated with mud which dried *en croute* to preserve their lanolin, producing wool so prized that English visitors would be taken to see the crusty ruminants as if they were a stop on the Grand Tour, as essential as any Michelangelo statue carved in marble. This fabric, with its threads finer than human hair, absorbed the indigo dye that enabled the colour known as navy blue, advertised in the eighteenth-century press as the patriotic shade to wear in wartime; an austere, deluxe colour, the same deep blue which adorned Beau Brummell's dandiacal body. Once milled, felted and pressed, the cloth lay so neat and tight that it needed no hems. The raw edges of this coat are just that: raw, still bearing the cut marks of the tailor's scissors, shaped out of his art with exquisite precision.

History takes over. Nelson steps out from Luzzardi's canvas, leaving a human hole in the canvas, to lie on the melamine table for our intimate examination. The effect is fixating, sensual; the cloth lies around the admiral's absent body as a fabric memory, all curves and flaps and seams and pleats. The tailors – Gieves of London, Meredith of Portsmouth – worked late into the night by candlelight focused through glass bulbs, filled with water rather than electricity, to create these miraculous constructions.

There's a naval strategy to this design: the sweep of the lapel, the rise of the collar, the arch of the pockets; they too contain codes. These are maritime coats, made for the sea, amphibiously adapted from the land and worn with white woollen breeches for warmth and ease of movement. They might be soaked with seawater, but such outfits were never washed. Shoulders would be sponged, breeches brightened with pipe clay; I daresay they'd have stood up of their own accord. The pleated skirts swung as Nelson walked, an echo of the extravagant attire of the macaronis of a generation before. Rather than set or follow fashion, military garb absorbed and slowed it down, incorporating it into its own masculine flash. Any other man covered in such fine cloth and gold lace – cuffs as weighed down with gold thread as any chunk

327

of gold chain – might be regarded as offensively flamboyant. No one could accuse Nelson of that, at least not to his stitched-up face. Yet the implicit swagger of those skirts combined with his torn face, his scars, his missing arm and eye to create an autofact as reconstructed as his vessel; an elegant man of war at home in a salon with his mistress, at a torchlit gothic party thrown by a noble pederast, or on a lurching deck, dealing death. As Amy says, such a costume civilised a hired killer. A murderer in the ballroom.

As we work through the three coats, I continue with my questions, as I would over the necropsy of a cetacean. What is this for, why is this there, what does that mean? Why all this gold thread and silver wire, sequins and shiny green foil, so brilliant and ersatz that they might have been confected from the wrappers of a tin of Quality Street? The three Nelsons laid before us are eloquent of his stature as well as his status: the narrow-cut shoulders, the nipped-in waist, the length of the coat are evidence of a short man, five foot six inches tall. I long to slip my arms in those slim sleeves and feel the skirts sway and bounce like a kilt; to be constrained by its tight chest and back; to be a hero, just for a minute or two. Yet for all the finery of these items, they are overshadowed by one last detail. In each coat, Nelson's redundant right sleeve – which goes unlined, in a random act of economy – ends in a thin silk loop, fixing the empty reminder of a former skirmish to his coat front; as much a badge of honour as any of the other awards sewn there.

Our audience with the admiral is over. As Amy and Louise fold the clothes like valets returning their master's attire, I notice a sequin fall out of one coat – a dull tarnished disc, like the scale of a fish or a lizard's eyelid, a bit of stardust. I consider licking my finger and dabbing it up as an illicit memento. Instead, I do my duty and point it out to Amy, who tucks it back into the cloth. The coats vanish back into the collection. The unit rolls to a close with a click, and we sign out of the building.

———

In old age, Melville's youthful trips to Greenwich and Portsmouth came back to haunt him. They were the holding places for his last story, as if he had been saving them up. His visits to those ports, which had precipitated the triumph and failure of his great white whale, rushed back like a late tide. With them they brought the body and soul of the Handsome Sailor, washed ashore at his feet.

Billy Budd is a work of poetry only pretending to be prose. It was written as an elegy from the end of Melville's life, long since spent on land as a customs inspector on the banks of New York's East River. It is a melancholy, beautiful tale, seen through his personal, and a greater, history; the momentous century he lived through, yet which in many ways passed him by. Melville was always at sea, in his head. You could see it in his eyes, as if the ocean pooled in them, as if his eyes had become the sea.

In his 1923 essay on Herman Melville, D.H. Lawrence saw the American as 'the greatest seer and poet of the sea', 'half a water animal ... a modern Viking'. He had, said Lawrence, 'the strange, uncanny magic of sea-creatures, and some of their repulsiveness. He isn't quite a land animal. There is something slithery about him. Something always half-seas-over. In his life they said he was mad – or crazy.' He was the outcast and the sensor, like Thomas Jerome Newton, Martin Eden, and Jay Gatsby, a lost, lone figure, all at sea. 'For with sheer physical vibrational sensitiveness, like a marvellous wireless-station, he registers the effects of the outer world ... of the isolated, far-driven soul, the soul which is now alone, without any real human contact.'

And Lawrence saw it in his gaze, his pale blue eyes that took in too much light. 'There is something curious about blue-eyed people ... something abstract, elemental,' he concluded. 'In blue eyes there is sun and rain and abstract, uncreate element, water, ice, air, space, but not humanity ... The man who came from the sea to live among men can stand it no longer ... The sea-born people, who can meet and mingle no longer: who turn away from life, to the abstract, to the elements: the sea receives her own.'

From his exile on East 26th Street, in a dark townhouse where his granddaughter would recall the fright of seeing a bust of Antinous covered by a veil to keep it from Manhattan dust, Melville escaped for Fire Island. He stayed there with his family in the Surf Hotel, which advertised itself as the only establishment at whose 'very doors you may revel in the sand or sea' and enjoy 'all the beneficial effects of the Ocean, without the discomforts of a sea voyage'. Melville worked on *Billy Budd* in his room overlooking the ocean: the same shore where Wilde had swum, where Thoreau searched for Margaret Fuller, and where Auden and Isherwood would dally, too; a fractured, queer coast, a halfway place like Provincetown, between here and Manhattan and the open Atlantic.

His memory was stirred not only by the sea, but by what had happened to his own cousin. As a junior officer in the US Navy, Guert Gansevoort had been party to the conviction of a young sailor, Philip Spencer, hanged for mutiny in 1842. Gansevoort had done his duty, but he was haunted by the episode for the rest of his life. Drawing on this family remembrance, Melville dived into his books like a library cormorant, reading Southey's biography of Nelson and accounts of notorious naval mutinies. Looking back

SURF HOTEL,

Fire Island Beach.

to an era that ended just before he was born, he invested the sadness and splendour of his life in the fate of a blue-eyed sailor. His story was simple, like Billy himself; but it had all the power of a parable.

1797. The Royal Navy, busily engaged in fighting the Napoleonic Wars, is threatened by the enemy within. There are mutinies at Portsmouth and the Nore on the Thames estuary, sparked by the spirit of the French Revolution. Anarchy is about to be imported to British shores. There's an apocalyptic vibration in the air.

William Budd, sailing from Bristol, is impressed in the Narrow Seas off the English coast. He is seized from his merchantman, *Rights of Man* (named after Thomas Paine's revolutionary text), and brought aboard HMS *Bellipotent*, commanded by Captain Edward Fairfax Vere. Billy, beloved of the ship from which he is taken, becomes the darling of the ship he joins. His besotted shipmates do his washing for him and darn his trousers; they all but swoon, and the carpenter makes him 'a pretty little chest of drawers', much as if they might set up home together. His perfection has only one flaw: a speech impediment which gets the better of him at moments of crisis. For all his beauty, Billy is as fated as the Ancient Mariner; his stammer is his albatross.

As his powerful allure is transferred to *Bellipotent* – from the civilian to the military – Billy's charm quells all quarrels. But his popularity is a challenge to John Claggart, the master-at-arms

– the ship's policeman. Affronted by a beauty he cannot possess, Claggart frames the Handsome Sailor as a scheming mutineer. He bribes an afterguardsman (the name is not coincidental) to meet Billy in the lee forechains, one of those secret places in the ship, and proposition him with sinful insurrection. Billy responds violently, stuttering, his honour outraged. But his innocence has been tainted. Confronted by Vere and accused by Claggart, Billy's defect becomes fatal. Asked by the captain if this claim is true, the sailor's stammer leaves him tongue-tied, unable to deny the charge. And as words fail him, Billy uses his fists. He lashes out at his accuser; Claggart falls to the floor, accidentally killed by the blow.

The drama of the story lies in the meeting of these three men, these three symbolic powers. Vere, an introspective, educated man, perpetually looks out to sea, as if he might find the answer to some unexpressed question there. He is respected by his crew as a man who will lead them to victory against the enemy. Yet he lacks Nelson's common touch; like Billy, whom he loves like all the rest, he is failed by his powers of communication. He cannot engage with his men, other than on an airy level – hence his nickname, Starry Vere, drawn by Melville from a poem by Andrew Marvell. Unlike *Moby-Dick*'s Starbuck, whose name implies the sky secured, Vere's underlines his aloofness. Claggart too is detached and foreign, with his strange 'amber' complexion and uncertain background: there are hints that the master-at-arms may be fleeing some episode in his past. He is neither officer nor seaman; a disappointed man, a manipulator, a villain: all these we read into his brutal name. Like Ahab, he is the personification of impotent rage. In the 1962 film of Melville's book, as watched by Newton in his apartment, Claggart declares, 'The sea is calm you said. Peaceful. Calm above, but below a world of gliding monsters preying on their fellows. Murderers, all of them. Only the strongest teeth survive. And who's to tell me it's any different here on board, or yonder on dry land?'

Filled with signs and wonders, told slowly as if he had all the time in the world, yet shortened where *Moby-Dick* was long,

Melville's tale takes on the tone of another last work, The Tempest. With its ritualistic sparseness, it too could be played silently, in the way Britten's operatic version of Billy's story replaces words with music. Characters become types: Vere is removed from the natural world, too philosophical and internalised to see it; Claggart stands over the darkness, at the edge of the abyss; Billy is beyond them both, a natural child, his only flaw echoed by his own alliterative, childish name which stumbles over itself. He is eager to please and to report for whatever duty his masters demand of him, unquestioning, like a loyal dog.

Not that there is any time left for questions. Starry Vere knows Billy to be innocent, but the king's justice requires retribution: 'Struck by an Angel of God. Yet the Angel must hang.' Hoisted from the yardarm by his own shipmates, Billy dies crying, 'God bless Captain Vere.'

At the same moment it chanced that the vapory fleece hanging low in the East was shot through with a soft glory as of the fleece of the Lamb of God seen in mystical vision, and simultaneously therewith, watched by the wedged mass of upturned faces, Billy ascended; and, ascending, took the full rose of the dawn.

Out of the silence that follows, a murmur of insurrection rises from the assembled crew – the very reaction this punishment sought to forestall. The swelling protest is arrested by the silver whistles of the officers, a sound as shrill as a sea hawk. Billy's virginal body, hanging as a 'pendant pearl from the yardarm-end', is unsullied by the ejaculation experienced by other executed men (while Starry Vere stands watching, 'erectly rigid'). He is then sewn into his own hammock and, weighed down with lead shot, tipped into the deep in a fulfilment of his song, an echo of Ariel's, and Jonah's: 'Fathoms down, fathoms, how I'll dream fast asleep . . . Roll me over fair! | I am sleepy, and the oozy weeds around me twist.'

Where Icarus fell into the water, burnt by the evening sun, the rose-tanned Billy rises with the morning sun, before being lowered to the deep. The sea hawks circle over the spot, so near the ship that the crew can hear the crack of their wings and the 'croaked requiem' of their cries. And as Ahab is dragged down by the whale, and Ishmael bobs up out of the whirling maelstrom, clinging to Queequeg's coffin, rising from the dead like Lazarus, Billy's innocent body is hoist to the sky then consigned to the sea, both condemned and resurrected.

Twentieth-century critics noted that Melville had read Matthew Arnold's *On the Study of Celtic Literature* and other accounts, and shaped his sailor out of Beli or Budd, the Celtic sun god of glorious death. Billy is a golden talisman found in the sea, miraculously un-tarnished. He is part god, part animal – the word victim signifies a beast for sacrifice – just as Lawrence, writing about *Moby-Dick*, saw Jesus the Redeemer as Cetus the Leviathan. 'Everything is for a term venerated in navies,' Melville writes at the end of *Billy Budd*. He tells us that the spar from which the Handsome Sailor hung is passed as a relic from ship to dockyard by sailors to whom 'a chip of it was as a piece of the Cross'.

There is a third and final death in this trinity. In an epilogue, Captain Vere is struck on deck by an enemy musketball. The ritual is complete. And as he lies dying, in an opiated stupor, Vere mumbles not 'Kiss me Hardy,' nor even 'God bless the King,' but 'Billy Budd, Billy Budd.' It is not an expression of remorse.

Lodged in the archives at Harvard is the manuscript of *Billy Budd*. It is as much-patched as Billy's trousers, its pages still pasted and pinned together, evidence that Melville revised his work ever more urgently, as if fearful that the story might grow to the length of *Moby-Dick*, yet knowing he had little time left in which to accomplish it. Racing across loose sheets of laid paper, Melville's open and slanting longhand – forever catching up with itself – allows for fewer than one hundred words to a page, even as it expands to a novella; while the date, added to the top of the first page – 'Friday Nov. 16. 1888. Began' – begs an ending, perhaps his own.

Melville was nearly seventy by the time he came to write *Billy Budd*. His eyesight, never strong, was failing, and he was ever more 'the occasional victim of optical delusion'. (When I meet his great-great grandson, Peter Gansevoort Whittemore, in New Bedford, shaking his Melvillean hand and looking into his pale blue eyes, he tells me that the family's only relic of the writer was a pair of his thick spectacles.) Perhaps that was what fed his imagination, those eyes that had seen too much. He was plagued by the memory of his son Malcolm, who had shot himself in his bedroom at East 26th Street with a pistol he kept under his pillow; his other son, Stanwix, had died of consumption in a San Francisco hotel.

In these tragedies of children who had predeceased him, this Daedalus held his invention to himself, unwilling to let it go. *Billy Budd*, as pure and provocative as it is, is suffused with regret and glory. It has a double power, since it was not published for forty years, long after its creator made careful corrections and pastings and pinnings and crossings-out, each scissored slip subtly swelling its reach. Alluding to the love of the crew of the *Rights of Man* for their Handsome Sailor, Melville notes that they would even 'darn ~~the seat of his~~ old trousers for him', a line which he amended, perhaps because it gave too much away. Yet he hardly reined in the homoeroticism of his hero, whose entrance is announced by the impressing officer, suitably impressed himself: 'Here he comes; and, by Jove – ~~look at him~~ – Apollo with his portmanteau!' The author's love of double-entendre had not left him.

And if Billy Budd is a kind of folk hero, all but stitched onto a scrap of cloth or carved into a bit of whale bone, there's an odd naïveté to his creator: the bawdy sailor turned autodidact like Shakespeare, for whom the whale ship was his Harvard and Yale. Melville's writing, even now, at the end of his fitful career, had an unconstrained innocence and a coded knowingness, one which was let loose yet made more mysterious by the sea – like Billy himself. His story's subtitle – 'An Inside Narrative' – admits as much, as if he were caught between Elizabeth Barrett Browning, whose works he avidly read, and Virginia Woolf, who avidly read his works; the same prophetic quality which would cause Lawrence to declare Melville 'a futurist long before futurism found paint'.

The revisions and replacements and parentheses, the attached flaps and sliced scraps of stray words and gathered phrases play with the past and the future like the starman's cut-ups, adding layer upon layer, harnessing a fleeting memory while achieving exactly the opposite effect. Supposedly taken from a broadside published in Portsmouth, Billy's story might as easily appear in *Lascar*, or a novel by Genet or Burroughs, in *The Tempest* or on Mr Newton's TV screens. There's something of its innocence in *The Great Gatsby*, published the year after *Billy Budd*, whose hero is called 'a son of God' and dies outstretched on his pneumatic mattress-coffin, floating in his pool, 'the holocaust complete'; while at the end of the century, in her film *Beau Travail*, Claire Denis would reset it in the Horn of Africa, where the sea becomes the desert and the sailors foreign legionnaires, performing a brutal, bare-chested ballet, choreographed to a soundtrack from Britten's opera.

As much as he covered his traces, Melville laid flagrant clues for future readers. If the published text – which he would never see – wasn't enough of a giveaway, the manuscript betrays its intentions in its written hand. For page after page he rhapsodises over the elusive Billy, spilling words in his direction. He lovingly evokes 'a lingering adolescent expression in the as yet smooth face all but feminine in purity of natural complexion', the sort of boy Owen met on English beaches; a boy 'cast in a mould peculiar to the finest physical examples of those Englishmen in whom the Saxon strain would seem not to partake of any Norman or ~~foreign~~ other admixture, he showed in face that ~~mild~~ humane look of reposeful good nature which the Greek sculptor in some instances gave to his heroic strong man, Hercules'. He lingers over Billy's body like a movie camera, 'the ear, small and shapely, the arch of the foot, the curve in mouth and nostril, even the indurated hand dyed to the orange-tawny of the toucan's bill, a hand telling alike of the halyards and tar bucket; but, above all, something in the mobile expression, and every chance attitude and movement'.

And he constantly underlines the phrase <u>the handsome sailor</u>. The effect is similar to the cinematic manuscript of *Frankenstein*, in which Shelley replaced his wife's description of the Creature as ~~handsome~~ with the word 'beautiful'.

Melville was making all these things up, even as he drew on Celtic myths and naval legends. Billy has no precedent; we do not know where he comes from, nor does his creator. When asked where he was born, Billy replies brightly, 'God knows, sir,' adding only that he was found as a baby 'in a <u>pretty silk-lined</u> basket hanging one morning from the knocker of a good man's door in Bristol'. The words 'pretty silk-lined' have been added afterwards, in an improbable detail reminiscent of Stephen Tennant's tough sailors and their unlikely interest in ribbons, while the officer's response is worthy of Wilde: 'Found, say you?'

Suffused as he is with strength and beauty, possessed of comeliness and power, Billy is primal and animal, and 'Like the animals, though no philosopher, he was, without knowing it, practically a fatalist.' And if no animal suspects its own mortality, neither does he. 'Of self-consciousness he seemed to have little or none, or about as much ~~of it~~ as we may reasonably impute to ~~the animal creation an intelligent mastiff~~ a dog of Saint Bernard's breed.' Billy is with the beasts; he is wild, but ends up in chains. Like a bird, 'he could not read, but he could sing, and like the illiterate nightingale, was sometimes the composer of his own song'. He stutters like a bird, too. He is the Lamb of God, an animal sacrifice. But he is also a natural force, a noble savage, a foundling: Melville compares Billy – known as Baby Budd – to Kaspar Hauser, a lost boy; but he could be Peter Pan or Mowgli. In early drafts, Melville seemed to indicate that his Handsome Sailor was black, a conflation of the Trafalgar veteran at Greenwich and another memorable sailor he'd seen in Liverpool on his first visit to England in 1839:

> The two ends of a gay silk handkerchief thrown loose about the neck danced upon the displayed ebony of his chest, in his ears were big hoops of gold, and a Highland bonnet with a tartan band

set off his shapely head. It was a hot noon in June; and his face, lustrous with perspiration, beamed with barbaric good humour. In jovial sallies right and left, his white teeth flashing into view, he rollicked along, the center of a company of his shipmates.

A black Billy would have been a wondrous cynosure, in the true meaning of the word, from the 'dog's tail' North Star, the star around which all the others revolve. But our Billy moves among his mates like another luminary, 'Aldebaran among the lesser lights of his constellation ... A superb figure, tossed up as by the horns of Taurus against the thunderous sky.' His heavenly body is the summation of his creator's desires. And whether golden Adonis or dark star, human or animal, what was certain was Billy's perfection, confirmed by a single fault, like Byron's club foot. 'Though our Handsome Sailor had as much of masculine beauty as one can expect anywhere to see; nevertheless ... there was just one thing amiss in him.' Billy's stammer belies his gender, yet he is made almost asexual by his innocent beauty.

In 1828, Robert Dickson, a ship's surgeon on HMS *Dryad* serving in the Mediterranean, noted a remarkable case in his logbook. He described eighteen-year-old Samuel Tapper, one of the able-bodied seamen, as having a 'brown complexion – an active and hardy lad'. But Tapper had a secret, hiding in plain sight. 'I had frequently requested to observe this lad (who is an excellent swimmer) bathe with the other boys,' Dickson wrote. 'Tapper's breasts so perfectly resemble those of a young woman of 18 or 19 that even the male genitals which are also perfect, do not fully remove the impression that the spectator is looking on a female.'

It was as if Darwin had fished a merman out of the Med; this intersexed sailor sporting like a dolphin, an Orlando in midtransition, or even some new species in the process of evolution. There is more than a little voyeurism in Dickson's observations. Having watched from afar as the boy bathed with his mates – who appeared to ignore Sam's 'curious formation' – Dickson gained access to this alien body when Tapper was brought into the sickbay. On close examination, the surgeon discovered that the young sailor's breasts were glandular, 'not at all resembling the fat mammille of boys'. Dickson deliberated, in an objective, Enlightenment manner on the cultural context for his interest, evoking Shelleyan images of marriages of male and female, and a magical ability to change shape. 'I have been chiefly moved to notice this case,' he wrote, 'having lately seen in the Royal Gallery at Florence the statue of an Hermaphrodite, (so called) perfectly resembling Tapper, in breasts and genitals.'

Melville was always romantically drawn to transformation. He saw Billy's imperfection as 'an organic hesitancy', an ambivalence in a story which, like Moby-Dick, is filled with strange undercurrents and digressions. But it was also by way of reassuring the reader that Billy's fate drew on hard fact: the threat of violent subversion against the leviathanic state and what the true rights of man might be.

The Great Mutiny of 1797 had shocked Britain with its 'unbridled and unbounded revolt', more menacing than all of Napoleon's armies of the Antichrist. The ordinary sailor had become the enemy within. Off Portsmouth's Spithead, the crew of Royal George rebelled, demanding better food – vegetables with their beef instead of flour – as well as pensions to Greenwich. These were denied. One admiral, Gardner, was so incensed by the petitions – including one requesting absolute pardon for their actions – that he 'seized one of the delegates by the collar, and swore that he would have them all hanged, together with every fifth man in the fleet'. The mutineers responded by hoisting the red flag. On London, Marlborough and Minotaur, sailors refused to go to sea; when they raised their guns, their officers opened fire, killing five men and seriously wounding six others.

By now the unrest had spread to the Thames. The men of HMS Sandwich seized control of their ship at the Nore anchorage, south of Sea Reach, the wild point at which the river became the sea, past which Melville himself would sail, unable to write his journal for the 'jar & motion' of the ship. The mutineers' ringleader was

Devon-born Richard Parker, an educated man who had previously challenged his captain to a duel. (He was also a man ominously christened, this Richard Parker, slipping in and out of history, from the fictional Richard Parker of Poe's *Narrative of Arthur Gordon Pym*, published in 1838, a cabin boy who is eaten by his shipwrecked mates, to the real-life teenaged Richard Parker of Southampton who met the same fate in 1889.)

Under Parker's brief leadership, the Nore insurrectionists made avowedly political statements. They demanded the dissolution of Parliament and immediate peace with France. Theirs was a maritime utopia, proposed on the same stretch of sea-river where Frank Harris had offered Wilde his freedom, the shore from which Oscar might have set sail for his own utopia, or where Shelley might have floated his anarchist fleet. Physically achieving what the poets had tried to do with words, the mutineers' Floating Republic was a direct challenge to the landbound state: they blockaded the venal capital, threatening its lifeblood of trade. The authorities responded by stopping their own food supply.

The mutiny failed. Parker tried to take the fleet to France, but his supporters deserted him. Arrested, tried, and sentenced to swing from the yardarm, Parker declined a white hood and jumped off before he could be hoisted, breaking his own neck. Far from becoming a martyr like Billy, his body was displayed in a tavern for a week to discourage further dissent.

But such grim tactics did not seem to be working. As the Nore and Spithead mutinies got under way, the insurrection spread to Parker's native Devon. In Torbay, two mutineers, William Lee and Thomas Preston, were hanged from the yardarm of *Royal Sovereign*, signing off their dramatic death letter, 'We who this morning are doomed to bid adieu to this World ... launched into the Gulph of Eternity.' Their bodies were put into coffins drilled with holes and sunk off Berry Head (only to be later retrieved by Brixham fishermen and respectfully buried on land). And at nearby Plymouth, the third centre of Britain's maritime power, two sailors on HMS *St George* were sentenced to death for sodomy. A deputation came to the quarterdeck to ask the captain, Shuldham Peard, to intercede on the condemned men's behalf. But Peard was warned by two men who slipped into his cabin that the crew were close to mutiny. He had four men court-martialled and hanged that Sunday, despite protests at the profanity. (Nelson said he would have hanged them on Christmas Day.) Ten years later in the same port, William Berry – Billy Berry – described as 'above six feet high, remarkably well made, and as fine and handsome a man as in the British Navy', was hanged for sodomy.

All these bodies paid the price extorted by the state. Melville made no direct reference to such injustices, but his work is full of them; nowhere more so than in the starry innocence of Billy Budd. In the body and soul of the Handsome Sailor, he looks beyond law and the ordinary world, to a place where we might all be set free.

The manuscript of Billy Budd was found, not in a basket but a bread box – along with Melville's note to himself, 'Stay true to the dreams of thy youth' – thirty years after its author's death in obscurity in New York. The book was dedicated to Melville's long-lost English friend Jack Chase, 'Wherever that great heart may now be, Here on Earth or harbored in Paradise'. And it was left to England to resurrect Billy, as if he were a twentieth-century boy. When it appeared in 1924, the story alerted D.H. Lawrence, Virginia Woolf, E.M. Forster and W.H. Auden to Melville's startlingly modern writing. In his collection Another Time – published in America in 1940, the same edition which included his hymn to Icarus, 'Musée des Beaux Arts' – Auden reimagined Melville. He saw him in those last years, his beard greying, his blue eyes failing, sailing into 'an extraordinary mildness'. Mindful of his own time, Auden – who owned a shack on Fire Island where, unbeknown to him, the Handsome Sailor had taken shape from the sea – wrote that 'Evil is unspectacular and always human,' while goodness 'has a name like Billy and is almost perfect | But wears a stammer like a decoration'.

A decade later, in 1950, Benjamin Britten's *Billy Budd* appeared during a new era of fear. In New England during the McCarthyite purges, Newton Arvin, the Scarlet Professor who taught Melville to Sylvia Plath at Smith College and whom Truman Capote, his lover, called 'my Harvard', would be prosecuted when images of naked men were found in his office. For Britten, as for Forster, his librettist, and Auden, their friend, Melville's writing was an eternal, subversive response, embodied in the otherness of the sea. His opera subtly altered Melville's story. Its oppressive setting and impressed men evoked the recent trauma of war and fears of a nuclear world in which terrible new weapons were detonated in remote oceans. As his crew toil and sing 'We're all of us lost forever on the endless sea,' so Starry Vere echoes their words with his own: 'I have been lost on the infinite sea.'

Falling and rising and falling again, they're all lost, these unmothered men and boys, abandoned to their fate like poor beautiful Billy, like Ishmael, like their blue-eyed creator.

All of them lost forever on the infinite sea.

STELLAMARIS

At dawn in Bantry Bay, I dawdle in the still green sea, pushing out from the pontoon, gliding over meadows of weed. It's a dreamy sensation made anxious only by the organisms with which I share the water. Every now and again I raise my head, looking out for the jellyfish that drift out of the dark like spectral umbrellas, gently opening and closing, luminous in the gloom.

This summer they're here in great number, brought by the Gulf Stream to Ireland's Atlantic coast. Locals tell me their appearance is good news, but large blooms of jellyfish often indicate disruptions elsewhere. Warming waters are sending their predators – such as turtles and ocean sunfish – further north; overfishing and reduced diversity mean that animals which should be eating jellyfish and their spawn are not there. Jellyfish populations expand to fill this gap, and in turn eat the fry of other fish. As fragile as they seem, they are great survivors, like cockroaches. They have outlived all five mass extinctions of the past; they've been around for six hundred and fifty million years. Harbingers of a new extinction, one day they may be all that's left in a simplified sea, the ghosts of what has been.

I had come down to the harbour as soon as I arrived, almost as though I was ready to leave again. It was early evening and the sun's heat was just beginning to abate. The sea was even calmer then, as if the day had flattened it down. I stood overlooking the water, wondering if I should get in. Everything felt deep and still. A man in a wetsuit appeared, walking down the boat slope with his children. He greeted me openly, as everyone does round here, and responded to my enquiry.

'Sure it's safe enough.'

His children splashed about at the edge of the slope, not quite ready to leave its security. He didn't seem eager to get in either. Then I noticed again what I'd noticed at first: his twisted hand, held at an awkward angle to his body like a dog's injured paw; and I thought, ah, he can't swim properly. I felt that pathetic ache, the sort of love imperfection demands. I had visions of him flapping about in the water on his side as a sunfish does, so directionless with its odd, floppy dorsal fin.

Moments later he dived in like an otter, more elegant in the sea than out of it.

Another man appeared out of his car, tugging on a wetsuit. He asked me to zip it up for him. As I yanked the teeth together tight across his shoulderblades, I thought of Robin Robertson's poem, in which a selkie shrugs off its skin like neoprene, joins a dance, then slips back into his hide at dawn, leaving with a 'famous grin'.

'That's me away.'

Slowly, I realised there were two or three wetsuited people in the water, gently working their way through it under the soft evening light. On this island, which I was visiting for the first time, having waited all my life to be invited here, swimming was a tradition. My usual tentative, if not secretive approach – expecting someone to tell me off, or ask, 'Is it cold?' – was unnecessary. I was among friends, fellow selkies.

Encouraged by their presence – as if their bodies made the sea safer – I swam out into the dark water, aware of the weeds as they swayed like slimy whips, each trailing tendril coated in white fur and insistent in its attempt to entangle my limbs. The sea was warm, and tasted only slightly salty. It felt old. The water of saints.

Then my hand brushed one of the jellyfish. I shuddered: it was like feeling a corpse in the water, the great fear of an out-of-hours swimmer. Uncertain of the underwater terrain, I soon swam back, and padded up the slope to join the others. One by one they were returning, all with the same story, their evening swims curtailed by venomous caresses, by tentacles delivering sly pink weals to bare faces and necks.

I realised I had to come to terms with these gloopy aliens that had parachuted in, falling up instead of down, as lazy as the days were long. I went back to the water morning and night with the tides, greedily storing up luxurious summer swims against the hard winter to come. I tried to forget about the jellyfish.

One evening I found another family there, a mother and her three children. The kids stood hesitantly on the pontoon, next to a sign which instructed us that swimming or diving from there was forbidden. Come on, I said, as they hovered over the edge, it's lovely. And braver than I pretended to be, I jumped in.

You remember some swims for no obvious reason. There's some conjunction of conditions, of spirit and intuition; the realisation that a place is ready for you, and you are ready for it. That moment, the surrender of no-going-back, the instant of transition, the leaving of one element for another. The sunny evening turned white and green. The rush of bubbles came up like a sheet and I caught a brief glimpse through my naked eyes of the blackness below before I bobbed back up like a bottle.

Then the kids leapt in too, hollering as they did so, and I thought how much I liked to share the water with children. How they don't muddy it with rational thoughts and worries. They are up there one minute, and down here the next. No assumptions. Just instinctive shrieks at the shock of the cold, followed by ecstatic shouts.

Unable to stay away, I was back there at dawn. In the complicit quiet of the morning, with no one about, I undressed on the pontoon and lowered myself in, sliding into the water so as not to disturb its gelatinous spirits. I was becoming reconciled to the jellyfish, even rather fond of their company; they too made the sea seem less lonely. The waves I created lapped the sides of moored boats as I wound my way in and out of the animals' paths. They moved mindlessly, mantles embedded with corneal flashes of orange and brown. Cloudy as cataracts, they peered unseeingly through the ocean's skin to the sky; I saw through them like lenses, down into the deep below. Trailing their bridal trains, they were ready to reward my bravery with a tingling stroke.

They stung me, again and again, each sting less painful for being expected, almost loving, the sensuous water made manifest. Maybe I was on the way to becoming a jellyfish myself, de-evolving, with all this time spent in the water; as if my spine and the other bones that held me upright on land might dissolve, leaving me blissfully at sea, little more than a bit of human zooplankton to be carried out to the open ocean, with no momentum of my own.

I like the common names of the sea, the way they gather by stories as much as by taxonomy. Jellyfish are no more fish than starfish or shellfish, but it suits us to think of them that way. Someone called these creatures compass jellyfish because their markings resemble a compass rose as seen through the glass dome of a ship's swinging gimbal. But given their wandering nature, it's hardly an apt name. *Chrysaora hysoscella* sounds more mythic – Chrysaor was the son of Poseidon and Medusa, 'the one with the golden armour', glowing from within. When he first encountered jellyfish on the Cape, Thoreau thought that they were 'a tender part of some marine monster, which the storm or some other foe had mangled'. Yet they are inescapably, intrinsically beautiful: to me they resemble elaborate Victorian puddings confected by Mrs Beeton, quivering dishes of aspic turned out of copper moulds for the delectation of well-dressed guests.

As I watched them at eye level they began to coalesce, summoned by some silent signal like clouds forming and re-forming; as much weather as animals, nebulous medusae. More and more emerged out of the darkness, their pearly colours complemented by another species, their fellow cnidarians: moon jellyfish, *Aurelia aurita*, canopies flushed with pale mauve and shot through with deeper purple-blue gonad rings, all but flashing with fluorescence, as though powered by an electric current running through the water. In their floral nothingness, they were flowers come to life: hallucinatory, floating out of the end of the world, anodyne yet venomous, born male, becoming female. Gliding by my side, powered by their expanding and contracting bells, they were barely there at all, composed as they are of ninety-eight per cent water, often ending up as sad puddles on the beach to be poked at by passing children whose parents pull them away, anxious that the evaporating blobs might yet retain the power to harm even when reduced to a spat-out wine gum in the sand. Yet these animals are complex products of evolution, their languid tentacles reacting to danger or seeking prey with an extraordinary speed that belies their jellyish nature. And nearer us than we know: they are our common ancestors, in whom nervous systems first evolved.

As successful as they are, these creatures cannot survive our scrutiny; and they only become more unreal in the exquisite models made by Leopold and Rudolf Blaschka. In 1853, Leopold, who came from a family of Bohemian glassmakers specialising in artificial eyes, was prescribed a sea voyage after suffering the loss of his wife and his father. Sailing to America, his ship became becalmed for two weeks off the Azores. Leopold looked down into the dark sea and saw 'a flashlike bundle of light beams, as if it is surrounded by thousands of sparks, that form true bundles of fire and of other bright lighting spots, and the seemingly mirrored stars'. This phosphorescence resolved itself into tiny, jellyfish-like siponospheres, whose cousins, Portuguese men o' war, I have also encountered in those waters, their lurid purple-frilled bladders dangling their venomous colonies like bulging varicose veins.

Sketching the invertebrates he saw, Leopold returned to recreate these creatures in the medium of which he was a master, passing on to his son Rudolf this passion for turning oceanic organisms into miniature Tiffany lamps. Together they read reports from the *Challenger* expedition – busy plumbing the world's oceans – and kept an aquarium in their home in Dresden, stocked with marine animals and plants from Trieste Zoological Station and the famous aquarium merchant, R.T. Smith in Weymouth, England, as well as specimens they gathered themselves.

The Blaschkas' extraordinary techniques died with them – even now, no one really knows how they made their delicate, impossibly twisted and shaped models – but their creations have survived, caught in time like insects in amber. My friend Mary once took me to the Harvard Museum of Natural History, promising something amazing. I walked into the gallery to find the Blaschkas' frozen plants and animals shimmering and glistening in rows under glass vitrines. They had been commuted into immortality, entombed in their own beauty.

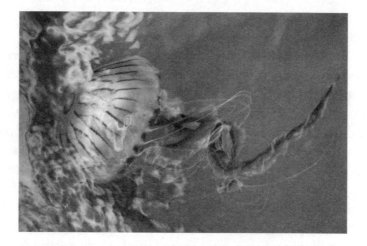

Behind me, Bantry was still asleep. On the shore, there were two pairs of children's shoes left behind at the water's edge.

Pulling myself back onto the pontoon, I dressed as the sun rose. Clouds slipped off the mountains and into the sea as though the land was breathing. I wheeled my bike up to the cemetery, its gravestones so shiny they looked as if they were polished every day. They marched up the hill, these neat slabs, tidier than the town's sprawling terraces, their black marble reflecting the dark water that they overlooked.

At the brow stood a cross, high above all the others. In an English churchyard such a monument would have commemorated a war. This memorial was dedicated, not to the lost sons of the Somme or Ypres, but to another generation.

$$+$$

TO MARK THE
FAMINE-PITS
OF
1846–8
MAY GOD GIVE REST
TO THE SOULS OF
THE FAITHFUL DEPARTED

The events of those few years happened only two lifetimes ago; my own ancestors left this island because of them. Their consequences are still felt, like the ripples around my body in the water. Held out to sea as a silent reproof, the stone cross is dumb. It cannot tell the real story, even now.

If this is the end of Europe, it is its beginning, too. Like Cape Cod, from which it is separated and joined by an ocean, a man might stand here and put all the continent behind him. And as with other places where the land runs out and the rest of the world rushes in – like Stornoway or Rotterdam or Lisbon or Galle – Bantry has the sadness and the beauty, the brutality and the abandonment of the sea. Its streets were once flooded thoroughfares that coursed as canals through the town. Now its harbour has been reclaimed as a car park, although on market day it is taken over by ducks and hens in feather-strewn cages, a dog wearing a hat, and a man with a brown paper bag over his head reciting his own poems. The public houses are still public houses, with their shelves of groceries and round tables, but the panelled front parlours, where the respectable ladies of the town once drank, are empty.

People have lived here for millennia. A medieval manuscript, *Lebor Gabála Érenn*, the book of the taking of Ireland, suggests that Bantry was the site of the island's first settlers, landing at a place which became known as Dún na mBárc, the Fortress of the Boats. In this account, Cessair, granddaughter of Noah, leads fifty maidens and three men to the western edge of the world forty days before the Flood in the hope that the waters will not reach them there. Only one, Fintán, survives the inundation. He turns into a salmon, then an eagle and a hawk, and lives for five thousand years before resuming human shape, to tell the story of Ireland. But in the mid-nineteenth century these mythic shores were invaded by a new species: *Phytophthora infestans*, a fungal pathogen brought over the same connecting sea, and its spots appeared on the potato's leaves as a sign of things to come. The first report was published in the *Dublin Evening Post*, 6 September 1845:

> We regret to learn that the blight of the potato crop, so much complained of in Belgium and several of the English counties has affected the crop, and that to a considerable extent, in our own immediate locality ... We are assured by a gentleman of vast experience that the injury sustained by potatoes from blight on his domain is very serious – that they are entirely unfit for use; and he suggests potatoes so injured should be immediately dug out for the use of the pigs.

With three million people dependent on potatoes to survive, disaster soon followed. They died in numbers so great that only individual cases can hint at the suffering of the whole. On 11 January 1847, *The Times* reported from West Cork with an opening line that might have come from a Dickens novel: 'The last accounts from this district are of a most dismal character.' Such dispatches would have been unacceptable if they had emerged from Hampshire or Devon; distance and disdain allowed them to exist, separated by a fatal sea. In the first year of the Famine, forty-two thousand tons of oats and eighty-five thousand tons of wheat and flour were exported from Ireland, while ministers in Whitehall declared trade had to be healthy to sustain the country during the crisis.

These were feudal scenes out of a new Dark Age, enacted at the outer limits of the industrial world. In Bantry, six inquests agreed that their subjects 'came to their deaths by starvation', including two-year-old Catherine Sheehan, who died at Christmas, having spent her last days eating only seaweed, 'part of which was produced by Dr M'Carthy, who held a post mortem examination on her body'. One mother and her three children were pulled from a freezing dyke; the post-mortem showed that the woman had not eaten for over twenty hours beforehand; she appeared to have drowned her children with her.

Other bodies were turned partly green from eating dock leaves, partly blue from cholera and dysentery. Dignified human beings began to resemble feral beasts, reduced to foraging in a landscape patrolled by packs of dogs scavenging on the dead.

Catalogue? Photograph? The *Illustrated London News* published eyewitness drawings by its artist, James Mahoney. He portrayed skin-and-bone beings – like the children of Want and Ignorance that challenged Scrooge's Malthusian enthusiasm – living in a land laid waste by some terrible, undeclared war. Evicted families set up shelters in the ruins of their own homes; known as scalpeens, they were little better than pits with canvas roofs. Others put up structures like whales' bones. Still more meagre were scalps, mere scrapes in the ground like the nests of shore birds.

Ireland was being ethnically cleansed, as New England had been rid of its Native Americans or Van Diemen's Land of its Aboriginal Australians. 'In such, or still more wretched abodes, burrowing as they can, the remnant of the population is hastening to an end, and after a few years will be as scarce nearly as the exterminated Indians, except the specimens that are carefully preserved in the workhouse.' Invoking the kind of images that Elizabeth Barrett Browning used in *Aurora Leigh*, the magazine saw it all as a by-product of industrialism, an act of imperial vivisection, 'a sort of Majendie experiment made on human beings – not on cats in an air-pump, or on rabbits with prussic acid'.

Imprisoned by the sea, at the edge of everything, the land allowed such scenes. With its liminal bogs and uprearing shores, its primitive rites and its infestations of spores, Ireland's insular fate was written in the water.

As a boy, I sensed a sort of Irishness in myself. I felt exiled in England – even though it was the only place I had ever lived. It was an otherness doubly bequeathed by being *Roman Catholic*, as if I weren't of my country at all.

St Patrick's, my primary school in the waterside suburb of Woolston, was physically and spiritually aligned around its Edwardian tin church, a somehow temporary building ready to be packed up should pagans overseas require its ark-like presence, or should the locals object to its Papist presence on English soil. Painted dark green, with a wooden spire and an interior fitfully lit by cheap stained glass, it was a colonial survival in a parish created in 1879 by the Irish chaplain to the nearby military hospital as a mission to this area of shipbuilders. Importing God into their lives, it looked a little like a ship itself, but was now used as our school hall where we attended assemblies and prize-givings. Across the playground stood another tin building, also dark green, imported from the field hospital at Netley. It is odd to think that my first classroom would have been familiar to Wilfred Owen.

Other classes were conducted in a cylindrical hut from the nineteen-forties; I thought its corrugated iron ribs were constructed to resist ordnance falling from the sky. In the summer, the grass grew tall over an underground air-raid shelter, a burrow built to protect schoolchildren from bombs. Now it housed our rufous and irascible caretaker, a cigarette perpetually in his mouth. Together, church and school occupied a block of their own, an island of faith in a sea of industry. At the bottom of the road lay the shipyards and a floating bridge, little changed since *Titanic* sailed from here ten years after our tin church was built, taking with it six parishioners who worked below its decks, men from Cork and Dublin.

Their apartness was underlined by the nickname for the area where I grew up and where I still live: Spike Island, a convict depot in Cork Harbour; a slur on Irish workers, who were seen as little better than criminals.

Next to our school was the new church, built in the nineteen-thirties. Over its entrance stood an eroding stone statue of St Patrick, after whom I was first named. Inside was a huge painting of him, dark and brown, a crozier in one hand, the other casting out the snakes slithering at his feet. On his feast day, fresh bunches of shamrock, stems wound with silver foil, appeared at the back of the church, miraculously imported from the Emerald Isle. They seemed seaweedy to me, grown in holy water. I never got to wear them; they were claimed by the black-haired ladies who mumbled the rosary through Mass as a never-ending chorus while the priest, faceless in crimson and yard-deep lace with his back to us, intoned Latin at an altar whose core was charged with the relic of a saint like a holy battery.

With its tall narrow windows of green and yellow glass, the interior had a watery light, like a giant aquarium. In an anteroom was the font in which my two sisters were drowned and reborn. Stone stoops contained that same irradiated water, which looked and felt and smelled like ordinary water but was, we knew, molecularly different. I waited for my turn to enter a cupboard where I knelt at the partition between me and absolution, confessing my sins through a grille as though to a cashier. Head bowed, the priest listened in the darkness, and sent me out forgiven in exchange for some penitential prayers. Ours were elemental rites: the anointing oil of chrism might have been whale oil as far as I knew, and the ashes scraped on my forehead at the beginning of Lent ground down from human bones. As we lined up to receive communion, the parquet floor yielded to the stiletto heels of young women, leaving fossil traces of their fashion in the herringbone pattern.

I may have had an overactive imagination, but nothing to a child is merely what it is, and I gave all these things other meanings. Perched on a leatherette kneeler, unsteady in my grey shorts and bare legs, I'd peer through praying fingers at the pair of double-height altarpieces made of stained glass set, not against the light, but in shallow niches either side of the nave, where the votive candles flickered on iron stands. They were unseeing windows, through which I might enter another world.

To the left rose the red-robed Christ, His Sacred Heart exposed in His holy chest and outlined in glittering tesserae of gold mosaic; He held one hand over us in a closed-finger gesture of blessing. To the right was the Virgin, enfolded in heavenly blue – more origami than gown – floating on a pale-green background imitating marble, two wavy blue lines at her feet symbolising the sea of which she was our guiding star. She hovered over the waves, calmer of storms, saviour of the drowned, star of the sea. It was she to whom we prayed, mourning and weeping in our vale of tears as we sheltered under her mantle. And in her bone-china hands she held the Christ Child, beatific in a purple tunic, suspended in front of His mother, an icon within an icon.

These altar images may have been made from stained glass, but nothing could be more pure. Their suspended figures and perfect doll-like heads surrounded by golden haloes bore little relationship to any reality. That is why I loved them and lived in fear of them. They could have come from Constantinople or another world; their open, almond-eyed faces and folded bodies seemed androgynous and eternal. I didn't know then that these larger-than-life-sized confections – which shimmered in the candlelight of winter afternoons as we recited the rites of Benediction, drugged by blue incense dragged across the altar in a swinging censer – were created after the smoke and fire of the Blitz had left the church a smouldering ruin, its ribcage roof reduced to blackened shards. Nor did I know that they were commissioned for the restored building from the Harry Clarke Studio in Dublin, as were our Stations of the Cross, fixed around the walls – fourteen square-framed scenes of Jesus's journey from judgement to crucifixion, depicted in painted glass jigsaw pieces held together in soft lead strips; fourteen stops of condemnation, torture and death, from station to station.

Christ had walked off the altar and into a cartoon strip. As Pilate washed his hands, Christ shouldered His cross; Veronica wiped His bloody face, leaving a ghostly image of suffering on the orange-brown cloth; He was stripped by the centurions and nailed to the wood and raised to the stormy stained-glass skies; and His body was laid in the sepulchre, as a lurid Palestinian sun set in the distance.

These glass pictures were a narrative shattered by trauma and put back together again. As the never-ending liturgy was intoned in unintelligible Latin I knelt in reverie, reflexively genuflecting, sinking in incense and hypnotised by Te Deums and heretical tedium, following the stations around the walls, heretically lusting over the soldier undressing Jesus, with his green cloak, shiny helmet, body-moulded breastplate and bare brown legs poking out of an armoured skirt, and fantasising that I might levitate into the air, defying gravity as the holiest saints could, hanging there like an angel to the amazement of the congregation.

These images anchored me to ritual itself. I would never escape them. I did not realise then what they meant for my body and soul: the Lamb of God, whose human sacrifice replaced animal sacrifice as He died for my sins. Like my blue notebook, their stories foretold my future. As if I'd already given up my own body, before I knew what I could do with it.

Henry Patrick Clarke was born in Dublin on St Patrick's Day in 1889. He was to grow up to be a reserved young man, his shyness at odds with the Beardsleyesque figures that he drew and which looked rather like him, with his huge dark eyes, angular face, slender hands and aesthetic air. After art school, he had joined the family firm and worked on church commissions. But his stained-glass, symbolist saints were decidedly secular and androgynous, almost opiated in their wide-eyed reveries; looking at them now, it is extraordinary that he was allowed to install them in university chapels, parish churches and convents, shedding their stupefying light on sacred sites. W.B. Yeats acclaimed Clarke, and the writer and mystic George Russell, 'Æ', called him 'one of the strangest geniuses of his time', who 'might have incarnated here from the dark side of the moon'.

Operating in the Celtic Twilight, Clarke's secular art grew ever more weird, from dreamy mermaids and staring revenants and ambiguous angels with red-gold hair, to Caliban figures with cloven hands and finned feet and etiolated, pale femmes that looked as though they'd grown in the dark, while dangling in the background, octopus-like phalluses peered through all-seeing eyes at the tips of their tentacles.

Caught between the fin-de-siècle and the new century, neither one thing nor the other, Clarke's figures tumble out of Wilde's *Salome*, the Russian Ballet and Keats and Shelley's fantasies of hermaphrodites and fauns, their gender and species undefined; equally, they might have emerged from one of Owen's poems. In their stained-glass incarnations, squeezed into gothic frames and psychedelic compositions as though the Abbot Suger had taken acid, they were born in the First World War and the Jazz Age, but their *lux nova* was medieval, evoking a pre-Reformation mystery. Their saints are elaborately faerie-like, folded in jewelled costume-coffins as stiff as the glass out of which they were wrought, contained by poisonous lead strips. Elaborate flowers interbreed with sea urchins on the ocean floor, clustered like the dizzying patterns in a painting by Klimt or the beads embedded in a Baccarat paperweight.

Ever since 1909, when he was twenty, Clarke had spent his summers on the remote Aran island of Inisheer, with 'nothing between him and America'. Staying on this 'very primitive place' with his friend and collaborator Austin Molloy, Harry wore the Sunday-best white felt suit and rawhide pampooties of the people of the sea; another friend, Seán Keating, depicted him reclining among the ruins of an ancient church like a languid monk. Painting and sketching by the shore each day, he became fascinated by the island's marine flora and fauna he saw in rock pools. Back in Dublin, he created a series of windows based on Keats's 'The Eve of St Agnes', painting lunettes – half moons – entwined with seaweed and jellyfish. He was working with light and water. His images were stained, acided and etched on glass made from kelp and sand from the sea, just as glass itself is a liquid formed of amorphous molecules through which we see light slowed down. Clarke's fabrications were environments as much as artefacts, as complex and fragile as the Blaschkas' creations. Indeed, he would have been familiar with the glassmakers' work: Dublin's museum of natural history owned more than five hundred of their models.

From sandy Atlantic shores via museum vitrines, those same medusa tentacles found their way into his illustrations for Poe's *Tales of Mystery and Imagination* and Coleridge's *Ancient Mariner*. Creating islands of his own, his work was set against nature as much as part of it; its blackness out of darkness dependent on the rising and setting sun. Filtering the sea light outside, his windows are inundated with gulf-warmed waters and studded with semiprecious reefs, while Christ's disciples slumber on the sea bed, pillowed by jellyfish and anemones.

In Dingle, the westernmost tip of Ireland, a peninsula that tapers into the Atlantic under a mountain named after Brendan the Navigator and in whose harbour I've just watched a lone, psychotic dolphin swimming endlessly from one boat to another, I wander through a decommissioned convent, its empty corridors smelling of institutional confinement. The sun forces through its chapel windows, loaded with Clarke's images. I wonder what the nuns made of these perfervid evocations of their faith, reimagined by a young man from Dublin who, stripped to a loincloth, had himself tied to the beams of his studio to pose as a living model for the Crucifixion.

With his puny body and bony ribs and his closed eyes he seems to have entered a transcendent, sacrificial state like one of his soporific saints, suspended in ecstasy, with flames all but issuing from his fingertips. He looks as emaciated as a famine victim or a shell-shocked soldier under observation, or the malnourished bodies he saw in Dublin's public baths. He was already a martyr. Weakened by the fumes of chemicals used in the production of his miraculous glass and by a near-fatal bicycle accident, he was sent to a sanatorium in Davos, Switzerland, in 1929 in an attempt to repair his consumptive lungs. He died two years later, at the age of forty-one.

And although Clarke had no physical hand in our suburban church – his son and daughter continued to run his studio, and it was they who supplied our stained glass – our Christ had Harry's dark eyes, his angular face, his unwrapped body. I look up at those images now and don't wonder at my childhood fantasies and all those creatures I drew in my book. All these saints were alternative stories for me, darker than any fairy tale. They held their fates before them – and before me too, in a world where war hung over my head like that nuclear sunset over Calvary. I had an overactive imagination, but what use is an underactive one? All that got you was a cheap suit and a briefcase. You might say they were the instruments of my oppression, but I thank God for glorious St Patrick, the saint of our isle, and for all the bleeding, martyred saints.

Bound up in my subversive faith, my intimations of Ireland were absorbed from a sense of obscure wrongs played out over the sea. I felt part of a separate and not entirely accepted caste. My school uniform was green; my mother knitted me an emerald-green jumper that I loved. I insisted that my eyes were green, even though people now tell me they're blue; perhaps they've turned blue from all the sea they have seen. Blue was the colour of conformity; I associated greenness with rebellion. My Irishness, as far as I suspected it, was another disguise; but I seemed to feel instinctively that I was an alien too.

My great-grandfather, Patrick James Moore, was born in 1856 in Blanchardstown, then a village six miles from the centre of Dublin. His father, Dennis, a blacksmith, and his mother, Rose Halpin, had been married in 1848, the year in which the Famine reached its peak. I imagine Patrick on his visits to the city, passing another young man in the street: Oscar Wilde, on his way to swim in Dublin Bay. And like Wilde, my great-grandfather would leave Ireland for England in the eighteen-seventies, albeit for wildly different reasons. Wilde would replace the hunger and disease of his homeland with decadent consumption as a gesture of unnatural defiance; my ancestor was driven out by the consequences of famine, and struggled to establish himself in a strange land. He moved to the port of Whitby, once famous for its whaling, where he worked as a seasonal fisherman along with the herring quines, gathering the fish before they too ran out. My grandfather, also named Dennis, was born in Whitby in 1886; the harbour lay at the end of their street. But economics forced the family to leave the sea for the mills of Bradford.

There, in 1914, Dennis married Josephine, the daughter of Michael Wall, a sail-maker from Limerick. He too had left Ireland in the wake of the Famine, from a port which witnessed the emigration of thousands. One newspaper reported people leaving Limerick 'as fast as sails can waft them from the shores of their fathers'.

These two Irish families came together in an English attempt at industrial utopia. Saltaire was Titus Salt's model mill town on the river Aire, outside Bradford; for Michael the sail-maker and Patrick the fisherman its name may have suggested the salt and air of their origin. Their children, my grandparents, were married in Shipley on 18 June 1914. Dennis and Josephine's wedding photograph might as well have been taken in Dublin, so Irish are its sitters. Framed by a bare painted backdrop, poised on wooden chairs set out on the studio rug, the new family are convened for a new century. At the centre sits my grandfather, a tailor-to-be in his elegant suit, pinned tie, boots shiny, moustache twirled and trim. He was little and bony, like me, with bright eyes. Into his memory I read my own Irishness.

Next to Dennis sits his handsome bride in a billow of lace, a twist of white flowers in her dark hair, a bouquet of red roses in her lap. Standing behind them is Bridget Wall, a veiled matriarch; her younger daughters, Rose Margaret and Kathleen, sit at either end. They look into the future. In a year's time Rose Margaret will be present at my father's birth, easing him out into this world in an upper room. Bound as much by otherness as by family ties, they all look confident enough; although my grandfather's knuckles are clenched around the brim of his hat.

Someone must have looked at this photograph and thought how life was better then, before the war, when three Irish brothers had married three Irish sisters. They had to become English in a country that looked on their old one with suspicion; plaster saints and the shame of famine were not wanted here. The Irish priest wrote out the register in Latin, converting my grandparents' names to *Denyius* and *Josephinus*, as though their Irishness were something to conceal. No one could imagine what was to come. In 1915 my father, Leonard Joseph, was born, a year before the Somme; his son would watch the starman on television; my nephew would see it all on his smartphone.

One evening, after dancing around the room with my father, Josephine went upstairs to bed and died of a heart attack. She was forty years old; she left behind five children. My father was just nineteen; he had to help bring up his younger siblings during the Depression. He would accompany his father on missions to give food to families, many of them Irish, so hungry that they snatched the bread out of his hands. He hardly ever spoke about those days, but he did recall a nest of rats running down the street, and a man whose body was found hanging in an outhouse on waste land.

Soon after, my father escaped the sooty streets of Bradford for the southern air of Southampton as surely as his grandparents had left Ireland behind. He was going back to the sea. He arrived at a station where the waves still lapped at the platform edge, although the port was busy driving its water away, reclaiming great stretches of land; the factory in which my father would work was built on that new earth, where giant cables rolled onto drums and onto waiting ships, tethering one country to another. He too had reinvented himself.

A photograph taken in the nineteen-thirties shows him as a young man, his hair slicked back, neat and handsome, standing on the coastguard lookout at Netley, close to the gates of the military hospital. He is posing to impress the photographer, my mother, whom he has just met and who knows this shore well; she was brought up here, walking this beach with her father.

In the background is a four-funnelled liner. Its silhouette is the same as Titanic's; the one was the ghost of the other. In the same way I would wear dead men's clothes, dead women's, too, as if I were an amalgam of my mother and father – which I am. The way they were then, what they aspired to be. Yearning for what we never had.

My father came alive by the sea. On day trips to Bournemouth he would exhort us to breathe deeply as if to get rid of the soot of those blackened houses up north where our aunts and uncles lived and in one of which, one dark morning, I watched a man step out onto the moon. But we never visited the country to which he owed his genes and his faith.

It has taken me this long to realise that my father was, to all intents and purposes, an Irishman, yet his connection to the island, and ours, had simply disappeared.

The city of Cork is Ireland's great exit point. From here sailed convict ships such as HMS *Java*, bound for New South Wales in 1833, with a cargo of two hundred transportees, among them twelve 'Whiteboys' from Kilkenny, violently opposed to Protestant tithes and guilty of swearing an illegal oath resolving to have a limb amputated rather than betray a brother. Their passage was recorded by Robert Dickson, the same surgeon who had placed the hermaphroditic Sam Tapper under observation; he noted in his journal that the Irish prisoners suffered far more than others during the voyage because they were so undernourished.

Transportees had no choice but to leave Ireland; the hungry and dispossessed had a choice, but not much of one. Thousands left from Cork's harbour at Cobh, their possessions parcelled up in brown paper, wearing their best, perhaps their only clothes. Melville recorded such scenes in *Redburn*, based on his first voyage to England in 1839. In Liverpool, Redburn sees a starving woman and 'two shrunken things like children' in a pavement vault, representatives of the refugees who lived in the streets and cellars, the sort of place that the young Heathcliff was found.

On Redburn's return journey, hundreds of migrants board his ship. The English passengers are protected by their twenty-guinea cabins 'from the barbarian incursions of the "*wild Irish*" emigrants', stowed away 'like bales of cotton, and packed like slaves in a slave-ship'. Even before they leave British waters, the refugees' spirits sink. Deceived by ship-owners about the length of their passage, they mistake the coast of their own island for their destination as they cross the Irish Sea: 'America must have seemed to them as a place just over a river.' One old man is only distracted from his search for land by dolphins riding the bow, shouting at them, 'Look, look, ye divils! look at the great pigs of the s'a!'

All this was so much invention. No such crowds boarded Melville's ship, and there was only one Irish name on its passenger list, Thomas Moore. Yet as a New Yorker, Melville was used to such sights; and to those who asked whether 'multitudes of foreign poor should be landed on our American shores', he replied, 'if they can get here, they had God's right to come; though they bring all Ireland and her miseries with them. For the whole world is the patrimony of the whole world.' Melville wrote Redburn in 1849, as refugees from the famine were arriving in New York in their thousands, many more dead than alive. They would resort to squats in the waste lands of Brooklyn, 'lying in the very heart of the city, and given over to hogs and cows, and to the squatter sovereigns who have erected wretched shanties upon it'.

From 1845 to 1855, two million migrants left Ireland for North America; it seemed as if the entire island was being transplanted across the Atlantic (just as its turf was exported to Cape Cod). In Black '47 alone, the year of the coffin ships, fifty thousand Irish died en route to America or shortly after they reached it. Vessels which had brought timber, tobacco or cotton to Britain were restocked with desperate people and overpacked to maximise profits on the return journey. The imperial British may have abolished the slave trade, but in this new fatal triangle, human ballast was dumped overboard and drowned, as Africans had been a generation before and are still drowning today. In scenes which might have been painted by Turner or filmed by CNN, other emigrants sidestepped the inevitable and threw themselves into 'the seething waters'.

The sea does not care. It never did. On his journey to the Cape in October 1849 – at the same time that Melville was sailing back to England – Thoreau came across the aftermath of another shipwreck. At Boston the Provincetown steamer was delayed by a violent storm; the same high seas had caused the brig St John from Galway, loaded with migrants, to wreck on the Grampus Rocks off Cohasset, across the bay from where Sylvia Plath would experience her own sulphurous storm as a child. Nearly one hundred and fifty Irish people had been drowned and were being washed ashore. The remains of the ship lay about in pieces; it was clear to Thoreau that it was rotten and rusty even before it foundered. Sightseers – drawn by a broadsheet handed around Boston, 'Death! one hundred and forty-five lives lost at Cohasset' – were milling about, gawping at the spectacle.

Thoreau – who appeared to relish his role as a deathly, transcendental beachcomber, as if he might find America's lost innocence there – watched as the bodies were recovered. 'I saw many marbled feet and matted heads . . . and one livid, swollen, and mangled body of a drowned girl, – who probably intended to go out to service in some American family, – to which some rags still adhered, with a string, half concealed by the flesh, about its swollen neck; the coiled-up wreck of a human hulk, gashed by the rocks or fishes, so that the bone and muscle were exposed, but quite bloodless, – merely red and white, – with wide-open and staring eyes, yet lustreless, dead-lights; or like the cabin windows of a stranded vessel, filled with sand.' When he found a large piece of the brig on the rocks, he was told that most of the victims lay beneath it.

Locals, as unconcerned as the figures in Brueghel's painting, were collecting seaweed washed up by the storm; to one old man the bodies were 'but other weeds which the tide cast up, but which were of no use to him'. Thoreau concluded, 'This shipwreck had not produced a visible vibration in the fabric of society'; the dead were so numerous, so strewn among the seaweed and their clothes so entangled with the wrack that, laid out in public, they lost their humanity. 'If this was the law of Nature, why waste any time in awe or pity?' he reasoned, ironically.

But one image stayed with him, like a nightmare: that of something white seen floating in the water days later, 'and found to be the body of a woman, which had risen in an upright position, whose white cap was blown back with the wind'. This vision darkened the beach for Thoreau. These desperate people had come in search of a new life, 'but before they could reach it, they emigrated to a newer world than ever Columbus dreamed of'.

On the other side of the Atlantic that same year, 1849, Victoria, dubbed the Famine Queen, arrived in Cork on her first visit to Ireland. Her royal gaze was carefully screened from the countryside's more terrible sights, from people whose hair stood on end as a side-effect of starvation, a precursor of some future dreaming. She declared Cork 'not at all like an English town', that it looked 'rather foreign'. Yet it was part of her empire, and even now its port – which was renamed Queenstown in her honour – retains a colonial air. Its Victorian terraces are dominated by the gothic spire of Pugin's cathedral, and out in the bay are the traces of naval installations which remained under British control until 1938.

The ferry chugs over the water, pop music crackling from tinny speakers. It is not a long trip. Spike Island soon looms up, greener than I expected, hanging in the harbour in the way Whale Island lies off Portsmouth, or Ellis Island off Manhattan. This was the famine prison, Ireland's Van Diemen's Land; the largest penal colony in the world.

The island has only recently become accessible; visitors are issued with safety warnings and informed about where and where not to go. Many of the buildings are decrepit, and on this sunny Monday morning we are told not to enter them. Peter and I pass crumbling grey barracks whose troops once manned the star-shaped fort. The grass has grown long and soft in high summer. Stonechats sing sweetly on concrete posts – although, as Peter points out, they are probably warding off interlopers to their territory. The tide is low, revealing a rocky shore, as good a barrier as any to men who could not swim.

Originally called Inis Pic – perhaps a reference to the Island of the Picts – its name was anglicised as Spike Island. This sliver of land, once a monastic settlement, had long held Ireland's unwanted; all the devils were here, too. Shakespeare may have seen Ireland as a model for Prospero's island – England's nearest, most troublesome colony as a plantation to be tamed and its wild people conquered – and Edmund Spenser, the author of The Faerie Queene, stationed in the county of Cork under Elizabeth's rule, thought the land should be subjugated and even consume itself, describing the victims of repression and famine creeping out of the woods and glens: 'they looked Anatomies [of] death, they spake like ghostes, crying out of their graves'. During Cromwell's campaigns in the sixteen-forties, thousands were transplanted from Spike Island like some invasive crop to the West Indies, to be superseded by enslaved Africans. In 1847, reacting to the onset of the famine, the island became a prison for men reduced to stealing food or defying unpayable rent. From August 1847 to August 1848, 2,698 sentences of transportation were meted out. Those unable to afford a migrant's passage stole to receive such sentences. In 1849, three seventeen-year-olds brought before Westport assizes accused of stealing hemp 'requested to be transported, as they had no means of living, and must do the same thing again'.

Two thousand men at a time were confined yet open to the harsh ocean and its weather, as though being readied for voyages to come. As near as it was to the land, the island appeared 'as isolated as if in the middle of the Atlantic'. Inmates were set to work in acts of useless labour; locals believed that they were made to carry buckets of water from one side of the island to the other, emptying and refilling the sea. Serious offenders were kept in darkened cells, converted from latrines. Those who worked outside wore caps with veils to conceal their faces as Wilde would do; many were made 'moonblind' by the whitewashed walls. Weakened by malnutrition, one in ten died of disease. Others, already traumatised by the loss of their families to the famine, went insane; the hopeless hanged themselves, or jumped off the cliff. Those who tried to swim away were recaptured and more heavily chained than before, as though to anchor them to the island. For some, their only resort was each other: in 1850, the Catholic chaplain, Fr Timothy Lyons, criticised for being too liberal, was admonished for failing to report 'indecent practices' between inmates in the privies.

On the far shore of the island a clump of trees conceals another decaying building, once the settlement's hospital.

'They call it Bleak House,' says Peter.

It has been entirely overtaken by undergrowth; to reach it we have to trample down chest-high briars and nettles. I remember such sites from my childhood; Victorian houses left empty, and yet not. The upper windows are boarded up, but at the rear, the out-houses stand blatantly exposed. Their roofs have fallen in on old toilets and baths, their glaring white ceramic obscenely spattered with rust and rot. The place is repellent. I use my camera as a kind of defence, quickly taking images before leaving it to its darkness. I feel as though I am visiting a dark version of my home, my Spike Island; a place where alternative histories were played out.

On the other side of the field a low stone wall encloses the graveyard. A team of archaeologists are digging in the dirt. Peter points out the dark stains in the exposed soil, each six feet long. The top layer is being carefully scraped away to reveal the remains of humans who long ago leaked into the earth. Twelve hundred convicts died on the island, but this small corral could hold two hundred at most. No one knows where the others lie.

Mara, a young American leading the gang of students clad in fluorescent vests and sweltering under the July sun, shows us where the wooden coffins were found, neatly carpentered and painted white by their fellow prisoners. Their contents left scant clues as to whom their bones belonged, beyond a few fragments of textiles. On one man's upper arm was a brass 'A', its significance a mystery. The only skeleton that may be identified – because of its size – is that of a boy who, according to the records, was just fourteen when he died here. The student chain gang digs on, in the same place where burial parties sweated two centuries ago. It is the unknown dead's fate to be constantly disinterred; these exhumed inmates will be commemorated in a service of remembrance, although no one remembers who they are.

Behind us, the parade ground is surrounded by three-storey Georgian barracks. In the nineteen-eighties the prison was re-opened to deal with a new crime wave of car thieves and substance abusers. When Peter was a boy, sailing in the bay, the island was out of bounds: convicts being ferried across would swear violently at him and his friends. In the hot summer of 1985 the inmates rioted and took over for a day, setting fire to one of the blocks. The burnt-out building is left eviscerated, its windows empty, its wooden floors gone. I peer into a tunnel-like space pierced by the sun from above. Trees have grown up through three storeys, reaching up to the light. Rubble and rubbish strews the floor. It looks like a war zone, a place for a strange meeting.

It's still early as I cycle up Rope Walk, high on the hills outside Bantry. Emerging from a green corridor of trees, I miss the turning at first, and backtrack to push my way through a farm gate, onto the open moor. At the brow, a slender post stands proud of the long grass, leaning to one side like something left behind. Only as I get closer does it resolve itself into a single shard of sandstone, about the height of a man. Isolated in the field, it draws me nearer. It hums with power.

There are carvings on its face, barely there at all; it might be a memorial to a dead horse from its grieving rider. But the Kilnaruane Stone is all that remains of a High Cross, dated to the eighth century. The horizontal wooden beam which once held its sacred status against the sky has long since gone, leaving only this gnomon, telling off the millennia like a sundial. The stone seems striated, stripped, as if it had spent centuries in the sea and having been cast ashore, got stuck in the ground like a bit of driftwood.

I try to make out the shapes, marks on a signpost to heaven, a wayside indicator to eternity. I need a map to read them.

An archaeologist's drawing shows a quartet of beasts, species unknown, although any reader of Revelations would recognise them as symbols of the Gospels: 'The first living creature like a lion, the second living creature like an ox, the third living creature with the face of a man, and the fourth living creature like a flying eagle ... each of them with six wings, are full of eyes all round and within.' Above is a sinuous, sinister form that could be a sea serpent or the writhing red dragon which menaced the woman clothed with the sun, the moon beneath her feet.

Across the valley is a holy spring, Lady's Well, where water gushes from beneath the feet of the Virgin, her toes resting on a crescent moon, stars and a snake. It is the same statue reproduced throughout the nineteenth and twentieth centuries – I remember it from May Day processions at school, when we carried it on a bier, singing to our star of the sea – the same statue you see from suburban Surrey to Sri Lanka, from the back streets of Brooklyn to the mountains of Mexico. Future historians may wonder at its meaning, just as they wonder at cryptic tattoos of the moon and stars on the bodies of transportees sent to Van Diemen's Land.

A mid-nineteenth-century antiquarian believes the Kilnaruane motifs to be one thing; a nineteen-forties archaeologist another; a contemporary researcher yet another. The shapes shift through history, even as I look at them, these saints and beasts setting off on their eternal journey. Cut into the stone cross beneath its synoptic creatures is a curragh, a skin boat rowed by four men, with another at the tiller, through a sea of crosses. It is the earliest representation of such a craft.

Some speculate that this scene shows Christ calming a tempest at sea, or His people navigating the storms of this world into the safe port of the next; His church as a ship, and its voyagers in the vault of heaven. Others see a similarity to the boat in which Brendan the Navigator sailed from these shores in the sixth century; hence the stone's local name, St Brendan's Cross, and the statue of Brendan in Bantry's town square, his arms outstretched over the prow of a boat. Just as Christianity first entered northern Europe through Ireland, preaching to seals on remote islands, so Brendan took the faith across the Atlantic. His adventures are recorded in a ninth-century manuscript contemporary with this stone: *Navigatio Sancti Brendani*, one of the immrama or holy tales of saints who set sail in search of isolation and peace. It may aspire to the power of a parable, but it reads like a medieval *Moby-Dick*.

As Brendan and his monks set out to find the Promised Land of the Saints, the Islands of the Blessed, they face mountains hurling rocks, griffins doing battle with dragons, and a rock on which they seek refuge and light a fire to celebrate Easter Mass, only to discover they have landed on the back of a whale. Brendan was unconcerned. (One chronicler claimed that the saint spent seven years on a whale's back: 'It was a difficult mode of piety.')

The sea was alive. And far from leaving them adrift, whales followed the faithful throughout their voyage, swimming around and under their boat, reassuring them of God's grace. It was for this service that Brendan became the patron saint of whales (oddly enough, at birth he was destined to be named Mobhi, until other signs intervened). On another holy day, the feast of St Peter, the apostle of the sea, a whole school of monsters appeared, attracted by Brendan's singing. His monks took fright anew, peering down into water that was terrifyingly transparent, as if they were looking into eternity itself.

> Sing lower, Master; or we shall be shipwrecked. For the water is so clear that we can see to the bottom, and we see innumerable fishes great and fierce, such as were never discovered to the human eye before, and if thou dost anger them with thy chanting, we shall perish.

At this Brendan rebuked his men; the Lord would deliver them from danger. 'What are ye afraid of?' he said, and sang louder than ever.

'And thereupon the monsters of the deep began
to rise on all sides, making merry for joy.'

This Christian Prospero had conjured up familiars to accompany him across the infinite sea. On the other side of the Kilnaruane stone is a knot of knitted serpents, which to some historians suggests the ancient sea god Manannan, often depicted as a sea monster. Perhaps he was the pagan creature Patrick cast out.

Below is the figure of a saint with his hands held outward in the *orans* attitude of open prayer; and underneath that, a pair of desert fathers, Anthony and Paul, sit at a table while a raven delivers their breakfast, the bread still in its beak. The raven abides here, in its western refuge. On other stones, notched with the Celtic alphabet of Ogam, these whalish clicks become the clonks and caws of the corvid, Brani, and a warrior named Brandgeni is a man born of a raven.

None of these images have definition any longer; they have been lost to dark time. It is eight hundred years since this cross stood in a wooden church, itself the shape of an upturned boat. Long after the worshippers had left the site was used as a cillíneach, a burial place for unbaptised children. Later it became a famine pit.

Standing by the stone on this lonely morning, I feel godless and godly. As if this human-high pillar were a petrified me. As if it had all come down to this rock, driven into an island. I had to wait to be asked here. As I leave, my place is taken by the black shapes of hooded crows, riding up and down with the wind.

That afternoon we sailed across the bay on Mark and Eoin's family boat, with Tara and Sinéad and newly-born Anne Marie, under low summer cloud that promised to disperse. There was a sense of imminence and potential to the day, of things waiting to begin, needing only some invisible cue.

Out of Bantry's harbour, rocky peninsulas rose on either side of us, jagged edges of land pointing out to the Atlantic. The sliver of Whiddy Island – named by the Norse Vod Iy, Holy Island – slipped past our starboard, its green slopes supporting oil silos.

'Apparently they hold enough emergency supplies to last Ireland two weeks,' said Mark. A rusting jetty marked the site of a terrible accident in 1979 when an oil tanker, *Betelgeuse*, exploded, and fifty men lost their lives.

We sailed on, beyond the island. Manx shearwaters swept over the waves; cormorants spluttered into flight as we drew near. The dark stubby dorsal of a harbour porpoise moved through the surface, rolling on its own axis.

Deftly, having sailed these waters since they were boys, the two young men steered the boat into an inlet, cajoling and persuading it, as if it were innately part of them. Then Mark stood up and with a serious look, one hand over his chest and the other held in the air, made me swear a solemn oath. He was about to take me to his favourite place, and I was not to divulge its location to another soul.

Slowly, we drifted into a narrow inlet overlooked by lush trees and enclosed by rocks on which seals lolled like sunbathers waiting for the sun to come out from behind a cloud. In 1934 Virginia Woolf visited this bay, and saw its soft light and stretches of virgin shore as the original land. It reminded her of her childhood holidays in Cornwall, and she imagined it was how the rest of England had been in Elizabethan times, when Orlando was a young man. She felt that here, life was receding. But renewing, too.

This was where the Celtic spirits were driven, westwards to the ocean – or to the other world. As Yeats, another kind of magician, wrote, 'the water, the water of the seas and of lakes and of mist and rain, has all but made the Irish after its image . . . We can make our minds so like still water that beings gather about us that they may see, it may be, their own images, and so live for a moment with a clearer, perhaps even with a fiercer life because of our quiet.'

Could water be haunted, homeopathically retaining the memory of what it has witnessed? Do all those shores remember me, as they remember those who went before? What trace do we leave? What have we done?

I looked down: the water was so still that you could hear any selkie sing.

It felt like home.

There was a tint to the afternoon – perhaps that was in the retrospect of Sinéad's snapped photograph – an extended summer longing, an intimation of autumn's slow recess. Tara, suckling six-week-old Anne Marie at her breast, told me how they'd sailed here last year during a heatwave, and were vaguely annoyed when another, smaller boat followed them. The lone occupant drew alongside, saying, without introduction, 'Lovely weather, great spot. There's a lot of people dying.' He meant that in the heat, swimmers unaccustomed to the sea had drowned.

Ignoring his warning, Mark and Eoin and I stood up on the side of the boat and jumped in.

The water was deep and dark, and the seals took to it too, rolling over on their ginger-spotted bellies and slipping into the sea, their puppyish heads bobbing just beyond us, peering at us with curiosity in their big black eyes. Jellyfish floated past while we trod water. Anything might have lain below us, down there. The rest of the world had ceased to exist.

Or rather, it all came down to this: a clear, cold, reflecting pool, languid with life and the sense of its continuance; of all the summers that had ever been and were yet to come.

I shivered from the water as we sailed back; the sun was just beginning to set. Mark lent me his Aran jumper, and asked me to take the tiller. I steered the boat inexpertly but steadily through the islands, back to Bantry.

The light was already falling over the town as we pulled into the harbour.

THESEATHATRAGEDNOMORE

Hanging upside-down, feet in the air, head pointing to the ocean floor, I listen. The sound is filling my body, drifting through the blue, out of the black. It's a song I've known almost all my life, but I'm hearing it for the first time.

Somewhere close by, he is singing. He has chosen this place in which to perform. This Mexican bay – one of the deepest in the Pacific, falling to three thousand feet – shelves to just one hundred and fifty feet here. He is using the thermoclines, which conduct sound five times more readily than air, to relay his song.

No human is quite sure how this animal creates these sounds. His is not a voice like ours; we have yet to locate the mechanism for his vibrato. He ought to be dumb, as far as we're concerned. To maintain such long notes without breathing, a whale must pass air up and down his windpipe, turning his sinuses – even his skull itself – into a giant instrument. It is the sound, as Roger Payne says, of the abyss; a sound the size of the sea. And it will be audible – tangible – to other whales for hundreds, perhaps thousands of miles; a humpback singing in the Caribbean may be heard by his fellow whales off the west coast of Ireland. His song sounds like a keening threnody to me; but to another whale, it is a serenade of lust.

He is singing, loud and longingly, for a mate.

When I first got in the water I floundered about, trying to locate the sound. It seemed to be all around me in the darkness. My lack of balance and a sense of panic sent me back to the boat. I felt as if I'd passed up an invitation.

Then I realised what I had to do. If the whale hung head-down, then so must I.

Taking a deep breath, I upend my body, my feet waggling at the surface like a seal's flippers. Weirdly, I have to reassure myself that the water is deep; that I don't have to worry about banging my head.

As I dangle there doing my best to mimic an animal many times my size, I briefly become a receiver in flesh and bone, a human hydrophone.

For the next two hours he sings; he's probably still singing now, through the darkness. His song changes constantly, from deep burbling passages that make me grin with their suggestive bass, to high-pitched staccato whistles as if he were impersonating a dolphin. Then, as he runs out of breath – like me – the sound spirals in short squeaks, signalling his return to the surface.

I pull my head above the water, gasping. As I do, I hear his plosive blow. I look over to see his back break the surface, obsidian-black against the Aztec mountains. It arches, glistening under the midday sun, vertebrae rippling as their own sierra, reflected in his blackness. Then he draws down his tail in one languid motion, and resumes his unfinished symphony. His repertoire is of such a range, of such a colour and complexity, that it sounds like a hundred other things. Sometimes it sounds like a fine violin, sometimes like a blown raspberry. Sometimes like a wet finger run over an inner tube, sometimes like a mournful elephant lost in the forest. And sometimes it sounds like me.

Using whatever air I have left in my lungs I try to turn it into a duet, pathetically imitating his profundo through my pigeon chest.

My ribs vibrate. Is he listening? His hearing is far in advance of my own; he feels me through his own bones, conducting sound through his jaws to his inner ears. I'm separated from the sea by the air in my ears; I hear through the changing pressure of the air. He is intimately connected to the element in which he swims; he feels its vibrations. He is huge. I am small. But we're the same.

He must hear me better than I hear myself. But nothing will change the course of his composition, certainly not the puny human hanging in the roof of his world. Perhaps he's laughing at me.

The sun bursts through the blue, its rays converging on infinity below. I hold my position for as long as I dare. I am utterly vulnerable, surrounded on all sides. This world is on his scale, not mine. By stepping off the boat and into his domain, I've given up my own. I've lost my soul to the sea.

He carries on singing, but I have to get out.

Back on the boat, we follow him through his sound. I hear it carry up to the surface and into the open air. Forty years ago, I never imagined it would be like this: such a public broadcast, out of his environment and into ours. The entire bay is a sounding board for his desires. The ocean's skin reverberates as a vast loudspeaker, a natural amplifier resonant with his call. If we first hear sound through amniotic fluid, then this sound might be borne on a golden disc, out to exoplanetary whales swimming through aerial oceans. Who isn't an alien? We're all lost on the infinite sea.

A whale's song alters each season, like fashion trends or musical styles. This cetacean transmission is evidence of a cultural exchange, 'unparalleled', as Ellen Garland and her fellow scientists note, 'in any other non-human animals'. Whales are the only creatures other than us whose evolution has been shaped by culture, by learned behaviour passed on from mother to child. In our male-dominated world, we are vain enough to believe all non-human song is directed solely at the means of reproduction. 'It is just like Man's vanity and impertinence to call an animal dumb because it is dumb to his dull perceptions,' Mark Twain wrote. We are not alone; we never were.

Why should an animal create such a complex sound? It seems an extravagant luxury. We suppose he is advertising his reproductive fitness, or using his songs as sonar to detect females.

His sounds are so deep in register that they may stimulate a potential mate from afar, bringing a female into oestrus; acting, in effect, as remote foreplay. But given his awareness and his culture, given what we know, and what we do not know, it may be that this whale is singing for himself as much as for other whales. If, as the French philosopher and artist Chris Herzfeld notes, birds sing on long after their songs have done their work, if dogs are excited 'by the tumult of the waves', and if great apes weave grass and elephants draw in the sand, why shouldn't whales sing for the love of their own song? Darwin was shocked by the peacock. We cannot comprehend such beauty beyond ourselves; we must burden it with other meaning.

For years I've watched these animals in other seas, although I wonder sometimes if I've ever seen them at all. When a blue whale raises her flukes against the Azorean sky, carving out of the air an exquisite shape beyond any human architecture, isn't it possible that she knows the power of her effect, the subtlety of its form and colour, the same flow and shade of the sea of which she is an intimate part? 'In no living thing are the lines of beauty more exquisitely defined,' as Melville wrote, 'the grandest sight to be seen in all nature . . . snatching at the highest heaven.'

And when a family of Sowerby's beaked whales appears in the early morning off those black shores, their strange dark shapes moving silently through the water, their subtle blows and antediluvian beaks breaking the calm surface to announce their presence; or when Risso's dolphins leap and spy-hop, so impossibly marked and scratched that they appear almost entirely white, like cetacean ghosts, in the way all whales are ghosts; or when a sperm whale appears out of the same sea, her body uniquely shaded in grey, a pale band around her belly splintering into shards towards her flukes like avant-garde haute couture, then spinning on her back to look at me binocularly from below and leaving me gasping behind my perspex mask – don't all these cetaceans, whose names seem to belong to humans, signal their own stories, their own sense of themselves, rising to adore their own gods?

I have no idea. In the ocean, all this is happening, all the time, as it always did.

Suddenly, our Mexican whale, who knows no borders, breaks right off our bow, rolling in the waves with another whale. I'd been sure that there was a second animal in the area; I could feel it. Our whale turns on his side, as the two males join together for a while in greeting, or some other intimate exchange.

He is the first whale of winter; his song will summon all the others to the bay. Later he will compete with other males for the favour of a mate. Fins armed with barnacles will tear like ferocious weapons; tubercles will be sliced off rostra in the mêlée. The female, much larger, will swim on regardless, until she has made her selection.

All this has been going on for millions of years. As we watch the males meet and part, Isabel, my guide, observes that whales have lived through many changes of climate while ancient storms broke over their heads. A whale bears witness to the past and the future because it so exceeds our own little lives. Whales live in another history that takes scant account of our own, predating and postdating us as they may do, spouting their frothed defiance to the skies. How can we presume to take a photograph of a whale, to capture the image of another species? I am the alien in their world. I'm what was left behind.

Here in my room, overlooking the ocean, I flop onto my bed, too excited to move, too dumb to speak, caught between day and night, between water and sky. Frigate birds fly by. All I can do is lie there in the heat, aligned to the sea outside and the fever inside of me, feeling my heart beat, staring at the ceiling. The rest of the day, landbound and lonely, is impossible to bear. After all this time they still have the power to unnerve me. Won't they ever leave me alone?

As the evening falls, somewhere out there he is still singing, if he ever stopped, a barnacled angel, bending sound. He sounds like me as I sigh in bed, unconsciously bemoaning my physical existence; as if I'd laboured too long in this body, weary of dragging my bones around in this gravity, and might reach out a flipper to turn myself over, sleepily twisted with oozy weeds.

Perhaps I could just let go of the world above – such a slight transition – and allow myself to fall, swaying gently from side to side, a last little dance before becoming something rich and strange. There we would lie in the cosy darkness, back in the never-ending night where I could hide and shelter, where no one could get at us ever again.

Back in my Cape attic, another storm rages, racking and rocking in the winter light, turning day into night. It cannot be defeated. It is brutal and beautiful, battering at the windows and tearing at the deck, fit to rip off every plank and tip me into the sea. It will not cease. It does not care. It blows and rains all day. I try to get in the water, but it picks me up and throws me out again, along with all the wrack and ruin.

I listen to the whale again, a long and low lament, dredging the ocean. And I wake hours later, in the dark, to hear the news.

I go out and walk the beach, waiting for the stars to fall. I can't do anything. Down the coast – I imagine – he lay in a white room, wondering if that was the way to go. It feels as though I came here by appointment, as if I'd written it down in my blue notebook, forty years ago. Everything falling away.

A single star shines in the morning sky. The window is a black mirror. I watch him on my black screen, seeing my face in his, an anchor drawn on his powdered cheek like a sailor from Amsterdam. I feel the energy pass through me like a discharged current. What does it take to feel that way, all over again? The feel of satin on my skin. Your eyes, at the centre of it all.

Dennis and Dory and I walk the winter dunes. The moss crumples under my feet. The sandhills go on forever. Dory shivers for me. She follows, wide-eyed, as I walk on my own, the third beside us, down to the shore. I sing out loud to the ocean and scratch his name in the sand. The waves soon wash it away.

Dory stands there, watching, as I take off my clothes and swim, like a dolphin, in the freezing sea.

... We are such stuff
As dreams are made on; and our little life
Is rounded with a sleep.

The Tempest, Act IV, Scene One

CREDITS

Editor: Nicholas Pearson. Copy editor: Robert Lacey. Text design
& layout: Richard Marston. Line illustrations: Joe Lyward. Cover
design: Julian Humphries. Picture research: Jordan Mulligan.
Publicity: Patrick Hargadon. Marketing: Tara Al Azzawi.
Representation: the late Gillon Aitken, Clare Alexander, Lisa Baker.

THANKS

Cape Cod, New England & New York: Pat de Groot, Dennis, Deborah and
Dory Minsky, John Waters, Mary Martin, the late Frank Schaefer,
Jen Bradley, Sacha Richter, Laura Ludwig and Stormy Mayo, John
and Marilyn Gullett, Elspeth Vevers, Elizabeth Bradfield, Jonathan
Sinaiko, Chris Busa, Jessica Strauss, Jo Hay, Sebastian Junger, Tom
Thompson, Albert Merola and James Balla; Mark Dalombo, Todd
Motta and John Conlon at the Dolphin Fleet; Robert Tarr Edmunds
Jr; Jim Bride; James Russell, Christina Colnett and Robert Rocha,
New Bedford Whaling Museum; John Bryant and the Melville
Society; Peter Gansevoort Whittemore, Jim Bride, Concord Free
Public Library, Andrew Delbanco, David M. Friedman, Mick Rock.
The Azores: Serge Viallelle and João Quaresma, Espaço Talassa.
The Netherlands: Jeroen Hoekendijk, Ellen Gallagher. *Mexico*:
Laura Logar, Isabel Cárdenas Oteiza; Alfredo T. Ortega, Centro
Universitaro de la Costa. *Catalunya*: Claudia Casanova and Joan Eloi
Roca, Ático de los Libros; Francesc Serés, the Faber Residency,
Olot. *Ireland*: Peter Wilson, Ann Wilson and Jim Wilson; Mark
Wickham, Tara Kennedy, Eoin Wickham and Sinéad Ní Bhroin;
Aengus O'Marcaigh, Barra Ó Donnbháin, Alicia St. Ledger, Paul
O'Regan. *Scotland and the Western Isles*: Roddy Murray, Ian Stephen,
Julie Brook; WDC Scottish Dolphin Centre; Scottish Seabird
Centre, North Berwick.

United Kingdom: Joe Lyward, Adam Low, Martin Rosenbaum, Jill Evans, James Norton; Andrew Sutton and Rachel Collingwood; Angela Cockayne, Alison Turnbull, Gareth Evans, Olivia Laing, Viktor Wynd, Iain Sinclair, Horatio Morpurgo, Jessica Sarah Rinland, Edward Sugden, Alex Farquharson, Volker Eichelmann, Marc Rees; Claire Doherty and Michael Prior at Situations; Tim Dee, Chris Watson, Duncan Minshull, Mark Cocker, Robert Macfarlane, Cillian Murphy, Tilda Swinton, Neil Tennant, Michael Bracewell, Brian Eno, Nicolas Roeg, Merlin Holland, Rupert Everett, Hugo Vickers, Paul Kildea, Andrew Motion; Peter Owen, Jane Potter and the Wilfred Owen Literary Estate; Stephen Hebron and Helen Gilio, The Bodleian Library; Susan Usher, English Faculty Library, Oxford; Hal Whitehead, Luke Rendell; Torquay Museum; Fr Claro Conde, Mary Hallett, Anna Eades, Katherine Anteney, Nick Moore, Clare Moore, Sam Goonetillake, Nigel Larcombe-Williams, Clare Goddard, Michael Holden, Pamela Ashurst and the late Ron Ashurst; Geoffrey Marsh, Victoria and Albert Museum; Louise Simkiss, Amy Miller, National Maritime Museum; Harriet Williams, Jane Fletcher, Mehta Bhavit, British Council; Damon Teagle, Millie Watts, National Oceanographic Centre; Dan Brown, Will May, Rebecca Smith, Stephanie Jones, Carole Burns, Matt Kerr, Karen Robson, Joel Found, University of Southampton. And my family and friends: Lawrence, Stephen, Christina and Katherine; Oliver, Cyrus, Harriet, Jacob, Lydia and Max; Mark, Ruth, Lilian and Freddie; and Tangle.

Philip Hoare, Easter 2017

SOURCES

www.4thEstate.co.uk/RISINGTIDEFALLINGSTARsources and at www.philiphoare.co.uk

IMAGES

Also by Philip Hoare

Serious Pleasures: The Life of Stephen Tennant
Noel Coward: A Biography
Wilde's Last Stand: Decadence, Conspiracy and the First World War
Spike Island: The Memory of a Military Hospital
England's Lost Eden: Adventures in a Victorian Utopia
Leviathan or, The Whale
The Sea Inside

First published in Great Britain in 2017 by
Fourth Estate
An imprint of HarperCollins*Publishers*
1 London Bridge Street
London SE1 9GF
www.4thestate.co.uk

Copyright © Philip Hoare 2017

1

The right of Philip Hoare to be identified as the author
of this work has been asserted by him in accordance
with the Copyright, Design and Patents Act 1988

A catalogue record for this book is available from the British Library

ISBN 978-0-00-813366-5 (hardback)
ISBN 978-0-00-813368-9 (trade paperback)

Designed and set in Quadraat by Richard Marston

Printed and bound in Great Britain by Clays Ltd, St Ives plc

MIX
Paper from
responsible sources
FSC® C007454

FSC is a non-profit international organisation established to promote the
responsible management of the world's forests. Products carrying the FSC
label are independently certified to assure consumers that they come
from forests that are managed to meet the social, economic and
ecological needs of present and future generations.

Find out more about HarperCollins and the environment at
www.harpercollins.co.uk/green